SCIENCE, TECHNOLOGY AND SOCIETY IN POSTWAR JAPAN

Japanese Studies
General Editor: Yoshio Sugimoto

Images of Japanese Society: *Ross E. Mouer and Yoshio Sugimoto*
An Intellectual History of Wartime Japan: *Shunsuke Tsurumi*
A Cultural History of Postwar Japan: *Shunsuke Tsurumi*
Beyond Computopia: *Tessa Morris-Suzuki*
Constructs for Understanding Japan: *Yoshio Sugimoto and Ross E. Mouer*
Japanese Models of Conflict Resolution: *S. N. Eisenstadt and Eyal Ben-Ari*

Forthcoming

Enterprise Unionism in Japan: *Hirosuke Kawanishi*
Group Psychology of the Japanese in Wartime: *Toshio Iritani*
Changing Japanese Suburbia: *Eyal Ben-Ari*
Rise of the Japanese Corporate System: *Koji Matsumoto*
Science, Technology and Society in Postwar Japan: *Shigeru Nakayama*
Transferred Development: Western Technology and the Industrialisation of Japan: *Ian Inkster*

SCIENCE, TECHNOLOGY AND SOCIETY IN POSTWAR JAPAN

Shigeru Nakayama

KEGAN PAUL INTERNATIONAL
London and New York

First Published in 1991 by
Kegan Paul International Ltd
PO Box 256, London WC1B 3SW, England

Distributed by
John Wiley & Sons Ltd
Southern Cross Trading Estate
1 Oldlands Way, Bognor Regis,
West Sussex, PO22 9SA, England

Routledge, Chapman & Hall Inc
29 West 35th Street
New York, NY 10001, USA

The publishers gratefully acknowledge the assistance of the Japan Foundation in the publication of this volume.

Printed in Great Britain by T. J. Press Ltd. Padstow, Cornwall

ISBN 0 7103 0428 5

British Library Cataloguing in Publication Data
Nakayama, Shigeru
Science, technology and society in postwar Japan.
1. Japan. Science. Technology, history
I. Title II. Series
306.450952

US Library of Congress Cataloging-in-Publication Data
Nakayama, Shigeru, 1928–
Science, technology and society in postwar Japan / Shigeru Nakayama.
242pp. 216 cm—(Japanese studies series)
Includes index.
1. Science–Japan. 2. Technology–Japan. 3. Science–Social aspects–Japan. 4. Technology–Social aspects–Japan. I. Title. II. Series: Japanese studies series (Kegan Paul International)
Q127.J3N27 1991
303.48′3′095209045–dc20 91-7998
CIP

Contents

Preface

Aim

The study of Japanese science and technology (especially technology) is a fashionable subject at the present time, and numerous English language works appear month by month claiming to explain the 'miracle' of the recent rise of Japanese technology. Most of these works are, however, superficial treatments of Japan's recent technological performance, seriously lacking in historical insight. This book is an attempt to introduce a critical examination of the mechanisms by which Japan has promoted science and technology by looking at its postwar historical development.

Western authors tend to generalize about Japan, their writing often reflecting the viewpoint of an outsider. On the other hand, some books written by Japanese authors which are available in the English language, remain superficial success stories, if not propaganda, of Japanese achievements, from which the Western reader is expected to learn some lessons. This is the first monograph by a Japanese author in the English language on the Japanese scientific community, looking at the impact of science and technology on Japanese society and vice versa.

Process

For several years I have headed a group research project, consisting of some fifty collaborators, on the subject of science, technology and society in postwar Japan, with the support of the Toyota Foundation. The outcome of this effort, a series of about twenty volumes in Japanese, will appear from 1991 until the end of this century. Prior to embarking on this project, I edited a book entitled *Nihon no gijutsuryoku* (*Japanese Technological Power*, Asahi, 1986), which discussed some of the issues to be raised in the series. Furthermore, I have published *Kagaku to shakai no gendaishi* (*Contemporary History of Science and Society*, Iwanami, 1981), which deals mainly with the postwar period up until the 1970's. By combining these works, and several other writings of mine in both Japanese and English, I have produced an English-language, single-volume outline of postwar Japanese science and technology in its social context. I have intentionally avoided using the word 'history' in the title as this book cannot possibly include all the ongoing historical investigations of the larger project. This volume consists mainly of my own personal observations and interpretations. It has been written with the intention of providing an outline structure for the forthcoming Japanese language series, as well as its English counterpart which will appear in the future as an historical work, and which will include specialized volumes on topics such as Japanese atomic energy and micro-electronics.

By that time, this volume will, hopefully, function as a single volume outline history of postwar Japanese science and technology. My argument is more based upon my impressions and reminiscences of the period which I have lived through rather than meticulous documentation. Later volumes in the project will provide the latter. Inevitably, my bold generalizations will invite debate and discussion. I would be happy being thought of as a provocative writer rather than an authoritarian one!

Acknowledgements

This project started in 1986 at Professor Yoshio Sugimoto's home in Melbourne, when he persuaded me to write a book for his series. Three years later, again under his auspices and his encouragement, I had an opportunity to spend eight months at La Trobe University with ample free time and facilities provided to concentrate on preparing the first draft. His secretary, Mary, kindly edited an early version of the draft of the manuscript.

I was also fortunate to have the opportunity to work with Morris Low, a promising young scholar of the history of Japanese science in the city of Melbourne at the same time. We met once each week and discussed the subject. He has edited various versions and also supervised the English corrections made by Liza Powderly of my first draft. During my stay in Australia, I was visited by Hitoshi Yoshioka, whose critical comments have always stimulated me. Dr. Bowen Dees has kindly provided me with information relating to the Occupation Period and has also assisted in editing a draft.

Toward the completion of the manuscript, Drs. Tessa Morris-Suzuki and Ian Inkster in Australia kindly provided editorial advice. My final thanks must go to two old friends of mine, Professors Nathan Sivin and William Cummings in the United States. Their invaluable advice and insightful comments led me to considerably revise parts of my original manuscript. Dr. Samuel L. Coleman gave me editorial help and useful comments and Mr. Morris Low has again helped me greatly at the very final stage of editing.

It would not have been possible to have completed such a wide-ranging work without the assistance of the people mentioned above.

December 6, 1990
Shigeru Nakayama

1 Introduction

Prelude

In the controversial events surrounding the promotion of nuclear reactors, which frequently occurred during the 1970s and 80s in Japan (as well as in most industrialized nations), the scientific profession has been divided into many sectors.

Those nuclear scientists, who worked at governmental laboratories such as the Atomic Energy Research Laboratory of the Science and Technology Agency in Japan, plan for the autonomous development of feeder reactors and accordingly write budget allocation proposals. The primary concern of their laboratories is to promote expensive nuclear research and development, irrespective of its economic feasibility. Their plan is promoted by the bureaucrats whose major achievements are measured by their success in securing maximum budget allocation and its accompanying power within the government sector, rather than any perceived long-range social benefits of the project.

Scientific personnel in private utility companies, such as Tokyo Denryoku (Tokyo Electric) have a slightly different stance from that of government scientists. They assess and adopt whatever system of energy-furnace as long as it proves to be economically agreeable. Although they are heavily influenced by governmental decisions on energy policy, they are in the final analysis profit-oriented and hence quite often prefer buying imported technology rather than the product of endogenous governmental development. In general, they welcome nuclear energy that is most centrally controllable by utility companies rather than decentralized alternative energy sources, such as solar energy.

Academic scientists, notably physicists in major Japanese universities, support in principle the peaceful application of nuclear energy but are quite critical regarding the hasty application of nuclear physics. They claim that more basic research, essentially more funding, is needed in the academic sector, before putting it into practical commercial use. Their warning of nuclear hazards often gets into technical details, upon which anxiety-driven local residents at reactor sites depend.

Finally, there are ecology-minded citizen groups, who oppose nuclear energy itself. Unlike those in the three sectors above-mentioned, they are the people who do not draw any direct advantage out of current scientific research and development. Freed from vested interest in its promotion, they take a fundamentalist stance in criticism of and opposition against the current practice of science and technology. One typical group is the 'Entropy society' which was organized in the mid-1980s in Tokyo. It was a gathering of activist citizens, in which some concerned scientists participated. Such scientists are only a very small minority in the existing scientific community; their voices not being well heard and treated lightly. But they have a wider audience amongst not only activists but also the general public who are not necessarily explicit in expressing their concern.

Four-sector approach

As I have shown in the above example, postwar science and technology has been an area where the interests of various sectors of society have clashed. Science and technology develops according to the dynamics of the relationship between the interests of the following four social sectors: academic, public, private and citizen (see the next section for the definition and explanation of the four sectors). The nature and characteristics of a nation's science and technology are determined by the nature of the equilibrium maintained among these sectors. We shall, throughout the present work, consciously analyse the benefits to each of the four sectors and treat issues according to their conflict and compatibility in relation to these four sectors.

It is generally believed that present-day science and technology is promoted via the cooperation (and conflict) of three sectors: academic, public and private. The citizen sector, on the other hand, is considered to be the outsider with respect to scientific and technological endeavour. Few works have dealt with the scientific activities of the citizen sector because of its low profile in comparison to the other three. In this work, we hope to advance an understanding of citizen participation in the assessment of science and technology by developing a conceptual framework of four, rather than three sectors.

Conceptual Apparatus Employed

In the preceding section, I introduced an example to distinguish several kinds of scientific activity and to highlight the involvement of various sectors in science. In the following, I shall define these activities from the viewpoint of the sociology of science.

In order to illustrate the special features of postwar Japanese scientific performance, it is no longer adequate to employ only the traditional taxonomy of science, such as science as distinguished from technology and to maintain a dichotomy between basic and applied sciences. I shall, therefore, introduce a conceptual apparatus which is employed throughout this study.

Indistinguishability of Science and Technology

Although the conventional distinction between science and technology still largely holds now, it would be misleading to adhere to it in its strict sense. Before World War II, when academic science dominated knowledge production, there was a clear distinction between science and technology. Academic science was strictly science (an activity concerned with obtaining objective knowledge of nature) and did not refer to technology (activities that were useful to some social sectors).

The postwar mainstream scientific activities (such as nuclear fusion projects and satellite programs) are heavily funded and assessed by sponsors (in public or private sectors) who stand outside the scientific community and tend to have little or no interest in or respect for science for its own sake, resulting in no sharp distinction between science and technology in their minds. Hence in the postwar period, science and technology have been inextricably combined so that distinctive terminology has become increasingly meaningless. In common English language usage, 'science and technology' used to be grammatically plural, while 'it' now has become singular.

Such scientific activity can now be called 'sponsored science'. I believe that, in the postwar world, the distinction between academic and sponsored science is far more significant than the traditional differentiation between science and technology.

Science Classified and Defined According to Assessors

The crucial factors differentiating the several types of scientific endeavor can be found in the social and sociological mechanism by which each is assessed and utilized. In other words, the question of for whom scientific research is addressed is the most important factor in shaping and defining the character of each scientific activity. Other distinctions such as research location, form of presentation and sources of funds are not necessarily essential and often remain derivative.

Academic Sector and Academic Science

Firstly, we can posit a sector or group of practitioners, scientists and technologists, who act as the promoters of and determinant force in the history of science and technology. Conventional internal historians of science and technology would probably reconstruct postwar history by simply drawing up a chronological list of the achievements of scientists. Such an internalistic approach was viable in the age of academic science before World War II, when most scientific achievements were attributed to talented individuals within the academic scientific community, without reference to any other social sectors. I refer to this sort of scientific activity as 'academic sector science' or simply 'academic science.'

In the case of 'academic science', assessment is conducted by fellow scientists. This tradition has been maintained since the establishment of modern science in the seventeenth century. It is a form of science which is based on the principle of publishing or making the results of research available to the scientific community.

Within the scientific community, members present their

research for debate, discussion and criticism by their colleagues. Their work is sometimes recognized by publication in the proceedings of a scientific society or its journal. Research is initiated out of individual interest and pursued for reasons of personal honour and distinction meted out through a referee system. Thus, even in technological fields, if a professor in engineering submits his work to an academic journal, he is considered to be an academic scientist and his work viewed as the academic science of engineering.

This mechanism has guaranteed and strengthened the universality and objectivity of modern science. Classical advocates of academic science attempt to promote science for its own sake, science as a self-justifying, value-free activity possessing an objectivity that transcends secular interests. Albert Einstein and Hideki Yukawa were probably representatives of the last generation to embody this mythology of academic science in an age when science has progressively become more influenced by outside forces. Particularly in the case of costly big academic science, public sponsorship is imperative. It is here where tensions emerge between the mechanisms for assessment and sponsorship. Such conflict between assessor and sponsor is, however, not explicit in public and private science.

The best location for conducting academic science in Japan, as well as in the United States, is generally thought to be universities[1], where the traditional ideal of academic freedom is supposed to be maintained. Scientists argue for the academic freedom necessary to advance this vision of science and the autonomy required to defend that freedom. Even today, these ideas continue to influence the values and standards of university men and women and are still used to promote research in the basic sciences.

Public Sector and Public Science

The main current of postwar science and technology has not been characterised by such academic scientists alone. Few would disagree that, apart from the academic sector, the outstanding feature of postwar science and technology has been the way in which the policies of bureaucrats-the public sector-have determined to a large extent the major direction in which science and technology has progressed, for the sake of national defence and welfare or simply the strengthening of bureaucratic power.

This trend was already evident from the 1930s when the tradition of German academia, which held the premier position in the world at the time, collapsed and scientists were alienated from the cosmopolitan values they had embraced and were forced to subordinate their values to those of the Nazis. Furthermore, under the government wartime mobilization of science, which included the Manhattan Project during World War II, scientists and technologists in all countries found themselves in roles subordinate to the political system. Governments were encouraged by the good results scientists had achieved during wartime and thus gave their full support to organized and planned mission-oriented research, such as the atomic bomb programs, the success of which guaranteed the continuation of this type of planned and organized research into the postwar era.

Throughout history, the government's contact point with science has been primarily focused on defence-related research for the purpose of improving national security. Other scientific activities can also be found in a modern state. Nationwide surveying of resources, the maintenance of national standards, the health care system and even industrial development are included in its agenda. These activities are mostly performed in the in-house public laboratories and hence we shall define it 'public sector science' or 'publicly-performed science.' In the postwar period, however, the government sponsorship was extended to academic and private sector; thus it could be called 'publicly-sponsored science'. The major direction of 'publicly-sponsored science' is set and its output assessed and appraised by bureaucrats. Since we tend to place more emphasis on assessors than performers, we define it 'public science'.

Private Sector and Private Science

Similarly, 'privately-sponsored science' as practised in the private sector is appraised by private enterprises in order to increase profits. Thus we name it 'private science'. This R & D is defined as developmental research guided by entrepreneurial principles governing the industrial production of commodities.[2] While the bulk of defence science and civil engineering falls in to the domain of public science, mechatronics (Japan-made English to denote a combination of mechanical engineering and electronics) belongs to private science.

Between the two world wars, American private research laboratories accounted for two-thirds of total national spending for research and development (R & D) in the U.S. The hallmark of postwar Japanese research and development is, as is now internationally recognized, the predominance of 'private science' over 'public science,'[3] for the sake of maximum corporate profit.

Sponsored Science, Technocratic Science

Public and private sciences have many common features which derive from the effect of evaluation by sponsors rather than academic peers. When I emphasize this common feature or try to include both public and private sciences in a single word, I define it as 'sponsored science.'

While sponsorship and assessment are not the same activity, they tend to be closely connected. Since activities which are not intended for public (or publicized) knowledge have no mechanism of assessment open to the outsider, it becomes necessary to assume that sponsorship is an indication of favourable assessment.

In actual practice, however, 'private science' is structurally bound together with 'public science' to form a complex, seemingly monolithic, establishment. In the U.S. this complex under the dominance of the Department of Defence is named the 'military-industrial complex', while in the Japanese case as it has been guided by the Ministry of International Trade and Industry (MITI) some people call it the 'bureaucratic industrial complex'.[4] We shall define this sector the 'technocratic sector and its science the 'technocratic science[5]'.

Technocracy is, in its simplest form, defined as 'rule by scientific and technical elites'[6]. It is, however, hardly conceivable for a handful of scientists and engineers to be able to rule a complex contemporary society. Besides, bureaucrats, financiers, business managers, and even military leaders cooperate with the core scientists and engineers to form an establishment and common ideology. In reality, technocratic scientists and engineers tend to be the second-rank advisers, and at most, the potential elites. For the purpose of analysis, however, it is simpler to refer to the small number of elite scientists and engineers as representing the upper echelons of power in the technocracy.

In seeking the most efficient way of reaching a preset goal, technocrats often prefer national interest to human rights, thus often causing conflict with the citizenry.

Citizen Sector and Service Science

As the technocratic state of affairs continues, people come to realize the necessity of restoring democracy to science. This leads to the emergence of a fourth sector, the concerned citizenry, or 'citizen sector'. The citizens may be defined as people who do not belong to either the academic, public, or private sectors described above; in other words, ordinary people who have no vested interest in and derive no direct profit from the advance of science and technology. Though they are not directly involved in scientific or technological work, many citizens believe that science and technology should be for the benefit of mankind.

Those scientific activities which directly contribute to the welfare of mankind, I refer to as 'citizen-sector science' or 'service science.'[7] A typical example of service science, that is, science which is directly assessed by the citizen sector, is health care. In spite of the recent 'industrialized' trend, a general physician's medical service is even today directly assessed by his patients. Such a service can be referred to as 'privately performed service science.'

Apart from defence, in a modern state, welfare and services have come to be assessed and evaluated directly by the citizen sector, such as in cases of the weather forecast service, environment protection, nutritional research, library science and so on; for that purpose inhouse laboratories and taskforces have been established, though a modern technocratically-oriented developmental state like Japan emphasizes service exclusively to industry as exemplified by MITI-affiliated laboratories.

Such government activity constitutes a grey area between 'service science' and 'public science'. It may be termed 'publicly-performed, publicly-funded service science,' if it is sufficiently assessed by the citizen sector; otherwise, it remains just 'public science'. Certain service science activities exist which are assessed at the local government level, where day-to-day contact between the government and local residents is maintained.

Compared with the other sectors, the citizen sector, relatively speaking, receives the least benefit from the progress of science and technology, its main role being that of a watchdog, to ensure that its interests are being represented in government policy. I include this type of activity also in the category of 'service science.'

'Service science' mainly involves assessment which is open to the public, but it differs from academic science inasmuch as it is not based on appraisal by research colleagues. Open scientific activities, other than those addressed exclusively to academic collegues in a highly stylized form, are thus considered to be that categorized into service science. Ordinary citizens, or more specifically, science journalists in the mass media, whose writings are directly addressed to and assessed by the citizen sector, and some concerned scientists who are supported by citizen organizations constitute the main bodies of appraisal.

In reality, in a technocratic society, most people are obedient to the line which a handful of technocrats devise. Accordingly, by default the majority is subject to the opinions of the 'active citizenry'. Since the 1970s, these politically active citizens have been increasingly prominent in expressing their concerns about various science and technology issues such as air and water pollution, nuclear reactors, the green-house effect and so on.

This type of activity has been called by some concerned scientists, 'science for the citizen', 'science for the people' or 'science by the people'. In an effort to avoid sectarian overtones, I have generalized the term to one of 'service science'.

Two Other Criteria for Classifying Science

1. Publicized and Privatized science

Technocratic science, which encompasses the private, public and often academic sectors, is sponsored and appraised by a technocratic establishment. Scientific information that has been produced by research is worthless unless it is owned and privatized by one of the sectors. The information, if publicized, can be used by anyone, even competitors, and will come to have little economic value. Therefore, the results of 'sponsored science' or 'technocratic science' are, as a rule, privately owned and used by sponsors or the technocratic establishment and its organizations. Such scientific information can only experience a deterioration in quality since it is not exposed to the full barrage of open scientific criticism[8]. Military research and the NASA space program are examples. If we define science from this perspective, we may term academic science as well as service science, as being 'publicized science', and technocratic science as 'privatized science' to use a term coined by Sir Solly Zuckerman[9].

2. Competition Mechanism

Competition is the incentive for such intellectual activites as scientific research and development. In academic science, competition is between individual researchers, but in sponsored science competition between countries and competition between corporations is the main motivation for funding R&D. When competition becomes intense, governments and corporations tend to place their own needs over those of the citizenry. In order to prevent such situations occurring, it may be desirable to make assessments on the basis of the fourth sector, the citizenry. Such assessments, however, can never become a powerful regulating force as the citizenry does not incorporate any competitionbased incentives.

Since the fourth sector does not possess the professional drive and profit incentive of the first three sectors and as it is often on the receiving end of the activities of the other three sectors, citizen movements, such as the anti-pollution campaign, must combine with journalism if they are to influence the tide of technocratic trends so that the damage inflicted on civilians does not exceed acceptable limits.

Table 1 and figure 1 to illustrate conceptual apparatus

science	academic	private	public	service
assessor	peer	sponsor	sponsor	citizenry
publishability	publicized	privatized	privatized	publicized
competition	individual	firm	state	none
performer	university	private lab.	public lab.	media
sponsor	pub. or priv.	private	public	none
utilizer	academic	technocratic	establishment	citizenry

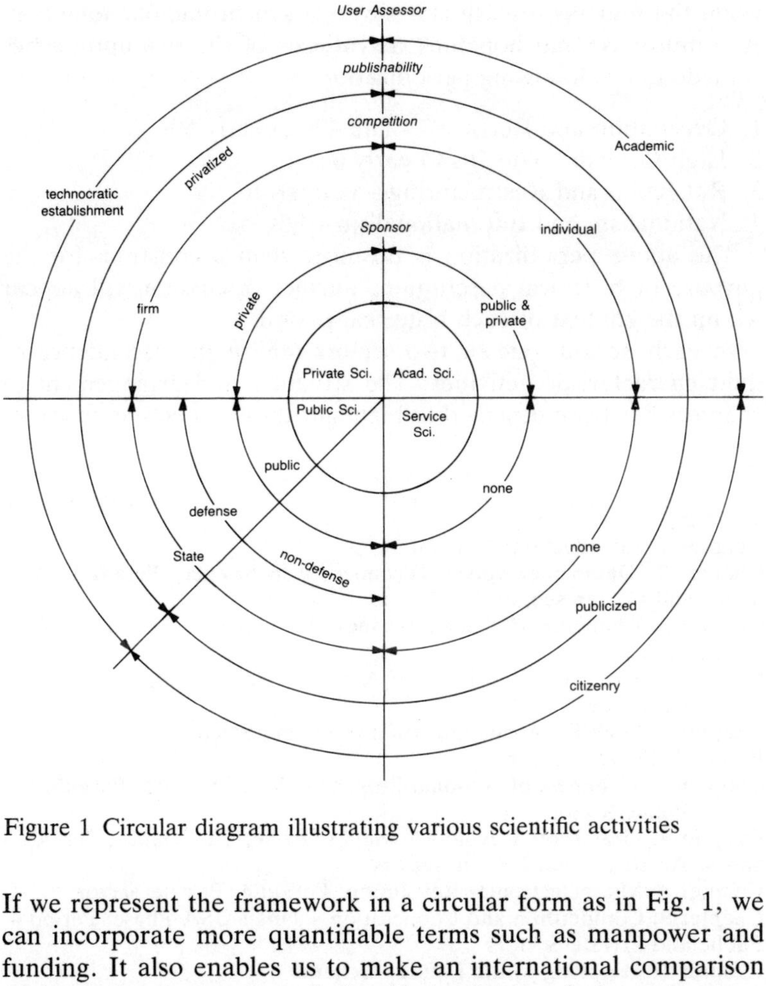

Figure 1 Circular diagram illustrating various scientific activities

If we represent the framework in a circular form as in Fig. 1, we can incorporate more quantifiable terms such as manpower and funding. It also enables us to make an international comparison of the national structure of scientific institutions in later chapters.

Periodization and Structure

In writing history, we constantly encounter the problem of a compromise between topical-vertical (diachronic, in F. de Saussure's term) and contemporaneous-horizontal (synchronic) approaches. In problem-oriented works such as this, we are tempted to take a diachronic approach to sectoral developments, while the foursector clash is basically a synchronic phenomenon. A compromise and hopefully a synthesis of the two approaches is made in the following periodization:

1. Occupation and Recovery – late 40s to early 50s
2. High Growth – late 50s to early 60s
3. Reflection and Restructuring – late 60s to 70s
4. Nationalism and Internationalism – 80s

The above periodization is no more than a construct for the purpose of historical description. Further discussion will appear within the context of each historical period.

In each period, one or two sectors realize the ascendence of their characteristic activities. The structure and arrangement of chapters has been designed to highlight these aspects as follows:

Table 2

Chapter 1: Introduction: General
Chapter 2: Democracy versus Technocracy in Science: Period 1. Academic and Citizen sectors
Chapter 3: Changing Models of Japanese Universities: Periods 1 and 2. Academic sector
Chapter 4: Expansion and Limit of Academic Science: Periods 2 and 3. Academic sector
Chapter 5: High Economic Growth and Private Science:
Period 2. Private Sectors
Chapter 6: Weakness of National Projects – Public Science: Periods 2, 3 and 4. Public Sector
Chapter 7: Grassroots Revolt – Possibility of Service Science: Periods 3 and 4. Academic and Citizen Sectors
Chapter 8: Microelectronics Revolution: Period 4. Private Sector
Chapter 9: Competition and Cooperation – Japan-USA Phase: Period 4. Public and Private Sectors
Chapter 10: Looking to the Future: General

Notes

1 In socialist countries, the location is the Academy of Science rather than universities, though the former are often inclined to public science rather than purely academic.

2 It encompasses a slightly broader range than that covered by the narrower term 'industrial science' which refers to work undertaken in research institutes within the company and includes projects delegated

to universities and think-tanks, as J. Ravetz called 'industrialized science'.

3 In contrast to the predominance of government sponsored research in major Western countries in the postwar period, Japanese private-sector research occupied two-thirds of total national expenditure for R & D in the 1960s and now more than eighty percent.

4 Johnson, *MITI*

5 I prefer to use the Japanese term 'Taiseika kagaku' for this kind of activity. In the absence of an adequate English equivalent, I tentatively use the expression 'technocratic science' in most cases, unless a specific usage is required, in which case I have used the following terms: 'establishment', 'incorporated', 'industrialized', 'government-sponsored', 'industrial', 'privately-sponsored', 'private' and 'public'.

6 Langdon Winner, *Autonomous technology* (MIT Press, 1977) p.135

7 As we define 'science' as the activity performed by scientists, we find it adequate not to call it 'citizen science' as the citizenry play its assessor's role, while scientific activity is performed by scientists.

8 Sir Solly Zuckerman, *Beyond the Ivory Tower* (1970) p.190.

9 *Ibid*. p.189.(or, in the words of Steven Rose, 'incorporated science.')

2 Democracy Versus Technocracy in Science

Immediately after World War II, academic and citizen sectors were outspoken in their discussions and debate about democracy and science (the 'rational spirit'). The public and private sectors quietly promoted economic recovery through technocratic means. In this 'tagmatch', where science is placed is the major issue looked at in this chapter.

The day of 15 August 1945 will long be remembered as the end of World War II. It is particularly indelibly branded in the minds of the Japanese, as it marks the turning point of every value system they held. The Japanese scientific profession is no exception.

The Japanese economic situation was a miserable one and hardly encouraging to scientific endeavor. Private laboratories were nearly closed and public science was subject to confusion due to the complete conversion from wartime to peacetime work. Academic scientists enjoyed relatively high morale in being at liberty to express their views. Neverthless, in embodying a new value system, it was a time when unprecedented innovative institutional reforms were attempted. Some of them are now sunk into oblivion or underevaluated in Whiggish hindsight.

In this chapter, I reconstruct some of the features of that period, witness the presence of forgotten ideology and ideals and furthermore trace its influence on the postwar development of Japanese science and technology.

Modernization in Japanese history can be interpreted as having begun with the initiatives of the Meiji government in the late nineteenth century, with the processes of modernization reaching the populace in a top-down fashion. Likewise, science and technology, which was synonymous with modernization in the eyes of the Japanese, was also introduced by the government sector, permeating the academic and industrial sectors and reaching the grassroots level.

Postwar Japan is demarcated by the American-led Allied Occupation and the concomitant introduction of the American version of democracy. In theory, democracy is disseminated first into the grassroots to build a firm base there, reaching up in a bottom-up fashion to influence the superstructure of political institutions. In the scientific profession, the bottom-up approach was also promoted as manifested by the scientists' movement to form the British Association of Advancement of Science in the nineteenth century as opposed to the old Royal Society.

However, as democracy is foreign-born and was in a sense imposed on postwar Japanese, its history should not be expected to follow the same pattern as endogenous democracy. In this chapter we shall examine the conflict between the top-down and bottom-up approaches in the history of scientific institutions.

Problematique

Is science for democracy or technocracy? If science is merely a superstructure, which depends upon a socio-political base, then what is democratic science and what is technocratic science?

Postwar Americans applied the simple yardstick of American democracy to gauge the degree of emancipation of the Japanese; democracy in science was no exception. The matter is not quite so simple, as we find a 'democratic' tradition continuing in the Japanese scientific community that is stronger than that of its American equivalent[1]. The Japanese scientific community has disapproved of the rise of hierarchical big science in favour of a more egalitarian allocation of research funds amongst its members. In the following we shall try to analyze the complex situation.

I. Postwar Democracy and Science (late 40s to early 50s)

During World War II, many democratic values were sacrificed in the name of urgency and efficiency. Democracy in science was forgotten in favour of preserving military secrets. Many scientists who had been caught up in the war effort were quick to demand a return to 'democracy' in science in the postwar period.

Japanese scientists, who were regarded by outsiders as belonging to a non-democratic feudalistic tradition, may well have felt themselves at a radical turning-point before and after 15 August, 1945, the VJ-Day. This first section deals with how they encountered 'imported' democracy during the ensuing Allied Occupation.

At the same time, we examine how and to what extent the characteristics of postwar Japanese science and technology originated in the science policy of the GHQ (General Headquarters), particularly the democratic and technocratic policy of the Science and Technology Division of the Economic and Science Section (ESS) and the Japanese reaction and response to it.

A. Military vs. Scientists during the Occupation

The immediate assignment of the Occupation Forces was the destruction of the wartime mobilization mechanism of the enemy and the appropriation of whatever was deemed useful for the winner's side. (The research results of human experimentation conducted by Unit 731 is one of the most atrocious cases.) Unlike the precedent of the case of Germany in the same year, the Japanese Occupation was virtually managed by a single nation, the USA, and hence, a more consistent policy was followed.

The military viewed science and technology as a part of its weaponry system. Immediately after the end of the war, the Occupation Forces could not overcome the fear that Japan might have preserved and concealed its fighting capability. Out of these fears came the desire to search for and destroy wartime scientific and technological facilities.

The most notorious and well-publicized incident was the destruction of four Japanese cyclotrons by the American Occupation forces. It was dramatically reported in newspapers that in September 1945, soldiers had destroyed what was the most advanced and precious scientific instruments in Japan. Certainly from the military point of view, cyclotrons were weapons to make atomic bombs, but scientists knew well enough that with the cyclotron alone it was impossible to produce bombs.

After the event, certain tensions arose in the USA between the military and the scientific community as to the best science policy for the Occupation. This was one manifestation of the power conflict occurring at the time between the two camps, and basically centred on who ultimately had the power to control the atomic bomb. Scientists claimed that atomic bombs are purely the product of the law of physics, which the military do not understand, while the military simply extended their traditional view of weaponry to include this latest product of physics.

Generally speaking, the military in the Occupation Forces who had close contact with the Japanese gradually came to understand the hardships facing the Japanese scientists immediately after the end of the War. This was in contrast to the attitude of the bureaucrats in Washington DC, who attempted to mechanically apply the directives which had been formulated before the actual Occupation was launched. In the case of the cyclotron incident, Dr. Yoshio Nishina made a plea to the GHQ for permission to use his instruments for medical and other purposes and it was granted after consultation with the American scientist-advisors. Permission was, however, suddenly cancelled and the order to destroy all of the Japanese cyclotrons was faithfully carried out by a couple of young officers of the Manhattan District. This action was severely criticized by the American scientific community, and their protests were reported by the mass media. Chinese scientists, who wanted the cyclotrons as part of war reparations, were most upset.[2]

At the time of the demolition of cyclotrons, the officer in charge found a discrepancy in the number of cyclotrons to be destroyed and the number he found at the site and entertained a doubt as to whether he should destroy remaining nuclear facilities. Accordingly, he sent a telegram to Washington to request the sending of a couple of physicists who would be able to judge what should be disposed of.[3]

In late December 1945, Drs. Gerald W. Fox and Harry Kelly of the Radiation Laboratory Project of MIT were visited by an army officer whose brief was to recruit them for the above mission. They were selected since 'they have not worked in the field of atomic physics and hence there are no misgivings of leaking atomic secrets.'[4]

In January, 1946, they arrived in Japan but at first the Occupation Forces did not know how to treat them. They had no assignment, and no job briefing was given. Quite at a loss, they almost decided to return to the United States. In the meantime, they were assigned to the Science and Technology section of ESS (Economic and Scientific Section). We have little knowledge of the planning behind the organization of ESS but the combination of Economics and Science reminds us of the prewar ideology of the New Dealers, using science for the reconstruction of the economy.

In this organisation, they were left to find their own role or job. The army was able to assign to them the job of military intelligence, especially the investigation of wartime military research and developments, and the unveiling of scientific weaponry such as the atomic bomb and death ray. They were not particularly interested in this type of intelligence work; as they had contacts with the Japanese scientific community, they came to know of the miserable economic conditions for scientific research on the one hand, but also how very hardy and capable Japanese scientific manpower was on the other. Based on these observations, Kelly concluded that it would be desirable to mobilize expert personnel for the economic reconstruction of impoverished Japan.[5]

Kelly entertained this ideal from the early part of 1946, much earlier than his colleagues in the same department. But this ideal was not anticipated in the early directives of the military occupation. Parallels can be drawn between Kelly and the so-called 'New-Dealers' in the Occupation bureaucracy, such as officials in the Natural Resources Section, who were instrumental in carrying out the postwar land reform of Japan, a project which could not be achieved at home but accomplished elsewhere only with the support of the all-powerful Occupation Forces. Kelly was, as he later described himself, the son of a poor coal miner, who could not indulge in pure theoretical research but had always been preoccupied with social usefulness. (Interview with H. Kelly.) Kelly's ideal achieved one expression in the creation of the Japan Science Council, the most democratic association of scientists which has ever existed, and which will be described in later sections.

B. 'Happy Marriage' of Democracy and Science: The Case of *Minka*

In January, 1946, the same month as Kelly arrived in Japan, a nationwide organization of Japanese scientists was created. It was the *Minshushugi kagakusha kyokai* (the Federation of Democratic Scientists, hereafter abbreviated as *Minka*).

Democracy and science are not necessarily rooted in the same traditions. Some may consider that both originated in classical Athenian culture, as they have a common origin in a uniquely Greek debating tradition.[7] Yet, evidence indicates that the two traditions developed independently in the course of time and that it is almost meaningless to look for common features and traces of an internal relationship between them. Before long, the problem of sponsorship arose, as in Hellenistic science. Not only in the Western tradition but also in the East, science was often supported and sponsored by absolute monarchs who had little interest in passing on its fruit to the grassroots level. Thus, because of its monarchistic flavour, science was neglected by the early American Congress.

'Democracy and science' was promoted as a political slogan in the Chinese May Fourth Movement of 1919. These two words were considered synonymous and became a symbol of opposition to the Confucian tradition. It is not very likely that the Japanese scientists after World War II were especially conscious of the Chinese precedent; yet, one common feature is evident. While the animosity of young Chinese intellectuals was directed toward the Confucian tradition, Japanese intellectuals and scientists saw a common enemy in the wartime militarism which had prevented the development of democracy and science from flourishing.

The stated aim of *Minka* to reconstruct a democratic Japan through the creation and dissemination of science attracted a wide audience in postwar Japan, and caught the interest of scientists, regardless of their individual political stances. The cadres of this organization were prewar Marxist activists, but compared to their prewar activities when they were few in number and organized in underground movements, the membership of the new organization was extended to practising scientists and engineers, active in every field.

The relationship between the Japanese Communist Party (abbreviated JCP) and the Occupation Forces was not at first as hostile as in the late Occupation period. In fact, the Occupation Forces right after the War were thought of, by the JCP, as their liberators from the oppression of Japanese militarism. The New Dealer faction of the Occupation Forces was instrumental in disbanding the *Zaibatsu* (big family-based financial houses) under a

Marxist-based strategy[8], and promoting the unionization of labor and the reform of land-ownership.

It was only the Communist Party which had a clearly formulated manifesto on science policy in 1946[9] (but no concrete strategy). Other political parties did not see science and technology as a concern warranting priority. The JCP condemned the old practice of Japanese science as 'feudalistic' and encouraged a new democratic way of science for the reconstruction of Japan. This ideal attracted many concerned scientists and engineers to the Communist Party and its associated *Minka*.

At the outset, they actively criticized and condemned the feudalistic and premodern characteristics of Japanese society in general and particularly the character of the prewar scientific community. They also were active in producing publications and helping to promote the popularization of 'democracy and science' in a leftist manner.

Minka was a purely voluntary association, which benefited from the conditions of that period offering scientists an opportunity to contribute to postwar democracy. In those days, scientists were seeking a new way of organizing themselves. *Minka* gave them the impression that what they were heading for was the only path for the future progress of science; that impression was commonly held by many of the young generation of scientists, including the present writer. At the height of activity in late 1949, the membership amounted to more than 10,000, a large number but an insignificant fraction of the whole scientific community.

1. Changing Relationship Between and Definition of 'Science and Democracy'

Democracy is, in its disinterested sense, nothing more than a particular form of government, namely the rule of the majority. When it is used as a political slogan against existing social institutions, it necessarily has the connotation of a moral and ideological character.

a. Science and Democracy as Liberators

Just like the Chinese precedent in the May Fourth Movement, the postwar Japanese slogan of democracy was popularly employed as an antonym of feudalistic morality. This changeover from feudalism to democracy was regarded by leftists as part of a historical course of inevitable development. For social scientists and *Minka* activists, the image of science was that of an inevitable historical law underlying society or scientific socialism in Engelsian fashion.

In the Japanese scientific world of natural science, medicine and engineering, this moral implication was often utilized when young dissenters attacked the old 'feudalistic', paternalistic, academic hierarchy. In contrast, 'democracy' was considered by them as having the moral character of a guardian of equality and classical liberty in the scientific community. The image of science here was largely that of academic science rather than of technocratic science.

Again in parallel with the case of the Chinese, science was viewed by proponents as virtually the same as democracy in its classical *Minka* enlightenment function as the liberator from feudalism.

b. Science and Democracy in Conflict

On the other hand, the theme of 'science and democracy', or 'control of science by parliamentary democracy', in the American version of political science as exemplified by Don K. Price's works, was never seriously taken up by the Japanese community. In contrast to the USA, in Japan the adversarial relationship between a technocratic administration and a popularly elected legislature has never existed.

The top government officials and representatives of the ruling political party have, in most of postwar Japanese politics, overlapped in terms of personnel so as to form a monolithic technocratic establishment. Hence, science and technology was not so much debated or criticized in Japanese representative democracy as in the USA, as can be seen by the activities of the Office of Technology Assessment of the U.S. Congress.

In early postwar Japan, science and technology was not a critical issue on the grassroots level apart from nuclear issues, as there was negligible involvement in military R&D. Since the 70s, however, pollution and environmental issues have drawn the attention of community residents, and nuclear reactor issues have now finally been brought to the centre of the debate between the citizenry and the establishment. At this point, democracy at the citizen level was redefined as a movement opposing the technocratic establishment. The Japanese approach consists of direct democracy resulting from the efforts of community activists rather than that realized through parliamentary representatives; the former places less reliance on parliamentary decisions in comparison with the American example.[10]

2. Academic Egalitarianism – Laboratory Democracy

As a result of Minka's condemnation of the old feudalistic scientific community and in the context of rapid growth in the strength of labor unions, a number of scientific laboratories experienced democracy movements right after the War. The first instance was at the Research Council (Kenkyu Kaigi) of the Toshiba Electronics Works (Toshiba denshi kogyosho). Started in November 1945, the movement proposed that the laboratory should be managed based on the participation of all the researchers, in contrast with the former practice of hierarchical management with command going one way from top managers down to researchers. This Council soon became a model for other laboratories to follow.

Another influential institutional paradigm was the Laboratory Constitution (Kyoshitsu kensho) initiated by a distinguished physicist, Shoichi Sakata at the Nagoya Imperial University. Sakata established a Laboratory Council (Kyoshitsu kaigi) system in June 1946.

During the war, Sakata and his followers enthusiastically studied J.D.Bernal's *Social Function of Science* (1939) and incorporated Bernal's views in their postwar proposals for the reconstruction of laboratories. Their aim was to overthrow the old hierarchical chair system and to base department management (including the appointment and recruiting of new members) on a decision-making mechanism in which votes are cast on the basis of the egalitarian participation of all the members of freely organized research groups (professors as well as assistant and graduate students having an equal right to vote).[11]

This system was welcomed and followed nationwide by those who sought a new organization of research, especially amongst young researchers in the physical sciences. There are, even now, the remnants of the Laboratory Council (Kyoshitsu kaigi) in many universities, especially in the natural sciences. Their origins can be traced to that period, even though the spirit of those days has all but gone. The system is quite idealistic by any account, but the problem of the conflict between elitism, which is inherent in any competitive scientific community, and the democracy of egalitarian participation, remains unrecognized and unsolved. This topic may be taken up later in more detail.

It might be pointed out further that this movement focused on democracy within scientific laboratories, but failed to formulate a link to the citizen level. We can say that the degree of contact with the grassroots level differs from one discipline to another, as does the concept of democracy. These differences were among the causes of the split amongst *Minka*'s branches, and the loss of solidarity and consequent decline of *Minka*. A notable incident was a split between the physics and geology branches of *Minka* on account of the establishment of a nuclear research laboratory; while physicists tried to persuade citizens that their work was safe, geologists stood on the side of the citizens under the banner of 'science for the people'. This will be treated again in the next chapter.

3. Egalitarian Forefront – Case of Grassroots Geology

Compared to physics, geological science was not a prestigious discipline in the postwar scientific community, but the role played by the geological branch of *Minka*, the *Chigaku dantai kenkyukai* (The Society for Group Research of Earth Sciences, abbreviated as *Chidanken*) founded in January 1947, is conceptually interesting, as it represents a form of grassroots approach to the sciences.[12]

Chidanken is not merely a scientific society but a crusading body out to propagate its ideology and methodology. Field geology can be carried out with the simplest equipment, a hammer, a clinometer and a pair of tough feet. As far as a particular locality or particular field is concerned, those whose labours involve endless walking and exhaustive observations, whether amateurs or semiprofessional geologists, know more than established top-notch elite scientists at a major university and laboratory.

In this respect a local historian or a local researcher is able to maintain his identity, pride and incentive for research. Likewise, the group research method, by organizing semi-professionals and students, has great merit, inasmuch as the exploration of mountains and fields requires as many human hands as possible. Thus, *Chidanken*'s democracy and egalitarianism have been brought into a close relationship at the organizational level; and conversely, in order to maintain the solidarity of its organization intact, *Chidanken* had to commit itself to a group methodology and the adoption of egalitarian democracy as its official philosophy.

Such being the case, the egalitarian principle of *Chidanken* stressed solidarity in group research whilst discouraging individual competition. Their system of fund allocation is so thoroughly based on the egalitarian principle that the society's leaders have neglected the procedure of the official screening committee for the Ministry of Education research funds and have distributed them equally to all members of the profession without any order of priority (mostly spent for the travel expenses of field research). This form of fund allocation fits nicely into the scheme of mass research, but in doing so the society has no chance of capital accumulation to purchase expensive instrumentation. As long as science is carried out in a labourintensive way, *Chidanken* provided a paradigm of research methodology, but as Japan recovered from her postwar misery, the trend toward capital-intensiveness invited problems.

While *Chidanken* still adhered to the postwar labour-intensive labour-collective approach, the budget for nuclear research in 1954, in response to President Eisenhowever's call of 'atoms for peace' in 1953, was epoch-making in the evolution of a capital-intensive research strategy and opened the path towards big science. Most scientists, notably physicists, were at the turning point from the poorly funded postwar approach, as exemplified by 'paper and pencil' theoretical physics to sponsored technocratic science of the establishment, which eventually led to the science and technology boom in the 60s. The labourintensive *Chidanken* could not cope with this new trend and its gradual decline ensued.

Such an egalitarian democratic spirit has been maintained in many other fields of science, though not so strictly as in *Chidanken*. Hoddeson has compared high energy establishments in the

US and Japan and found that the Japanese are so overdemocratic that the manner of fund allocation aims to make everybody happy but not too happy, while the American competes with a few candidates and a final winner gets everything.[13] In Japan this practice might have contributed to the suppression of the elitism that produces the numerous Nobel Laureates and hierarchical big science enterprises characteristic of major Western nations, notably the USA. While recent authors who emphasize the value of group commitment and conformity of the Japanese factory workers in quality control of industrial products would like to attribute the above characteristics to the inherent Japanese national character, I would say that it is rooted in a tradition which began in the postwar egalitarian democracy, in which everybody had to share a very limited amount of resources at a time of extreme economic hardship.

A point of departure from postwar democracy was the system of excellence inherent in the competition-minded scientific community. Academic science, since its beginnings in the seventeenth century, was mainly promoted as a game played on the basis of individual competition. As the generation of new ideas is a highly personal matter, it is hardly likely that intellectual fertility will emerge from mass discussion and consensus under the majority rule. Because of the egalitarian ideology of *Chidanken*, some outstanding members, notably Akio Miyashiro, were purged out of the organization. Their outspokenness and critical attitude toward an established theory often led them to step out of the consensus and doctrine of *Chidanken.* Some found in this point even a sign of totalitarianism, which inhibits any deviation from the central norm of the organization. *Chidanken* avoided the long dominance of a dictatorial leader, by adopting a constitution that only young scholars less than forty years old are eligible for the directorship. Their 'democratic' majority rule has, however, only contributed to the dominance of conformists to their pre-established norm.

The ideal type of research that *Chidanken* claimed, namely mass-organized research with egalitarian democracy as its principle, is appropriate for research such as precise geological map-making that requires an assiduous 'human sea' strategy (*jinkai senjutsu*, the strategy of throwing waves of men into action)'. In the late 60s, China proclaimed a new research method of science with mass participation, *jinkai senjutsu*, under the auspices of the

Cultural Revolution. The anti-pollution citizen's movement of the citizenry planned an exhaustive search for pollution in the early 70s in Japan via the mass mobilization of amateur researchers. Both these methods were anticipated by *Chidanken*, but by the late 60s and early 70s they had almost been banished from the ideological forefront, losing the pioneering vigour that they possessed in the early 50s.

Creation of the Japan Science Council

The major achievement of Harry Kelly's ideals was the creation of the Japan Science Council.[14] Kelly had the ambition to achieve something entirely new which would be parallel to the accomplishments of the Occupation New Dealers. What he found in postwar Japan was economic catastrophe, and he came to believe that science and technology should play a key role in Japan's recovery. Although he had a background in academic physics research, Kelly suppressed academic science and its symbolic institution, the age-honoured Imperial Academy of Japan, in favour of applied science that was to be employed for the immediate objectives of economic recovery.

In June, 1946, Kelly organized a number of Japanese liaison associations. These were private groups centered around him to provide advice for the reorganization of Japan's research structure. Liaison was, however provided with no authority and criticized as being an undemocratic organ. To offset these complaints, the more publicly visible Preparatory Committee was created. This group acted as a supervisory board overseeing the election of members to the Scientific Research Organization Renewal Committee, ensuring that the election of members was conducted democratically and that the elections were representative of the Japanese scientific community.[15]

Kelly was, of course, a firm believer in democracy and supported the laboratory democracy of young Japanese scientists and engineers, but in other respects he was closer to leading technocrats, such as Seiji Kaya[16] who were efficient and effective in pursuing the reform of Japanese academia. Whether Kelly's choices in science policy during the Occupation contributed to the later dominance of the technocratic model is a matter for further investigation. In his mind and in the eyes of the New Dealers, the choice between democracy or technocracy was not clear. At least he was more cautious in admitting the leftist line of the JCP affiliation than his fellow New Dealers. When he was approached by leftists, he encouraged the laboratory democracy they pursued but left them to work on it further with the Renewal Committee. Since the committee was dominated by technocrats and established academics, the *Minka* affiliated members were dismissed and effectively excluded from the Committee's decision-making.

Inside the Committee, opinions were split into two groups: the technocrat group, who were working in departments of the natural sciences and technology, tried to promote the opinions of scientists; and the low profile group, composed mainly of academic social scientists who, because of their painful wartime experiences, wished to stay out of politics and preserve their academic freedom. Finally, a compromise was made and the drafts of the Japan Science Council Law, as well as rules for the election of delegates were completed on July 10, 1948. Under this Law, the views of various groups were taken into consideration, including that of the leftist *Minka*, which fought to extend membership to cover those working in industry and the private sector.[17]

The draft was sent to the GHQ for approval. Kelly and his group modified it to limit the power of the JSC over the administration, fearing leftist domination of parliamentary decision-making. This occurred at the time Japan's shortlived Socialist government was in power.[18]

As the Renewal Committee was organised and its membership selected by the Ministry of Education, Japanese mainstream journalists often severely criticised the Committee's technocratic and authoritarian character. Commentators were all suspicious that the delegate membership would be monopolized by the staff of the University of Tokyo, the most prestigious government university and the institution which produced most of Japan's high level bureaucrats.

In the first election, however, many of the Tokyo University faculty members and even members of the Renewal Committee were defeated. On the other hand, more than ten percent of the leftist candidates were elected, in spite of their generally low status in academia. Candidates from provincial colleges and the private sector were also elected. On the whole, democratization was well under way.

C. Democracy vs. Technocracy in Academia – Fate of JSC

The Japan Science Council is a unique organization. It is an elected body from all the members of the Japanese scientific community, the qualification for membership being the authorship of at least one article in an academic journal or the recommendation of a recognized learned society. There may be similar associations, like the British BAAS or American AAAS, in which council members are elected from the entire scientific community. What makes the difference is the semi-official makeup of the Japan Science Council, as it is entrusted with allocating the government research budget. I have not heard of a similar association even in socialist countries. In the latter, the academy of science is, in practice, a government organization, but the new members are selected by old members, a practice which is found in most of the national academies of science of the Western world. It is exactly concerning this point where conflict arose between democracy for grassroot scientists and the technocracy of government favour.

From the outset, in the Japan Science Council, the tensions between the technocratic delegates and *Minka* delegates was obvious. The former were more pragmatic, while the latter tended to be more ideologically-bound.

The former group was establishment oriented, closer to leading elites in government administration and the ruling political party, while the latter was an anti-establishment group based on a tightly organised supporting body at the grassroots level of the scientific community, somewhat like a trade union, with ties to the Japan Communist Party. These two camps had virtually nothing in common. For the former group, the Japan Science Council was only one of many channels that may function as a conduit for planning and strategy, while the latter group viewed it as a stronghold of postwar democracy and leftist activities.

1. STAC (Science, Technology and Administration Council)

Coalition and compromise was achieved from time to time in the early stage of the history of the Japan Science Council, but in the meantime, the technocrat group found JSC inefficient for achieving its aims. At the time of founding the Japan Science Council in 1949 they created the Science, Technology and Administration Council (STAC), an organization associated with the JSC, in order to liaise with various administrative offices. STAC was chaired by the Prime Minister; its membership consisting of top bureaucrats of ministries and scientists recommended by JSC.

STAC had, as a rule, monthly meetings, but the government was quite unresponsive to it, as it was not quite the right time to accomplish the technocrat's dream. After 1954, the function of STAC was transferred to the more powerful Science and Technology Council (STC), the advisory body to STA (abbreviation for the Science and Technology Agency, which will be discussed after Chapter 4), to which the technocrat group transferred their base from the problematic JSC.

D. The Decline of *Minka* and JSC Apathy Toward Politicization

During the Occupation period (1945–1952), *Minka* and its progressive scientists protested about the indifferent attitude of the Japanese conservative government to the promotion of science. It was the government which played a major role in the promotion of R & D, inviting and incorporating scientists into their camp but in a technocratic way, which is considerably different from *Minka's* conception of science. This difference was, however, not keenly noticed by the orthodox JCP affiliated scientists nor by the average member of the scientific community. *Minka's* accusation turned out to be pointless as they were unable to formulate a clear and distinct outlook of science in accordance with the changing circumstances.[19]

At the same time, along with the development of a cold-war mentality within the Occupation Forces, the red purge tended to harden the attitude of the *Minka* group towards the conservative establishment. In March 1952, *Minka* put forward a new concept of *kokuminteki kagaku* (literally, 'national science', with an undertone of 'science for the people'), with the intention of opposing the Americanization and colonization of Japanese science, which was likely to follow the signing of the Peace Treaty, which was to be concluded without Soviet participation. There were certain followers among engineers who opposed the introduction of American technology out of fear that their professional position might be jeopardized. However, this strategy was not taken up by the scientific community at large. The majority quietly retreated from *Minka's* version of democracy and assimilated themselves into the technocratic regime.

As *Minka's* new concept of 'national science' or 'nationalistic science' was more a political slogan than a scientific methodology, many members were puzzled by its implications for their daily work; some internationalists were troubled that such a nationalistic appeal should be made, when bitter wartime memories were still lingering in their minds. In the context of international politics, however, the Korean War and American McCarthyism must have caused apprehension amongst leftist intellectuals.

In one word, 'national science' was a *narodniki* (back to the people) movement, created by the science and technology section of the Japan Communist Party, aimed at attaining nationwide support for their political position by disseminating scientific knowledge at the grassroots level. It was carried out in a topdown (scientists to people) manner. The extent of its success is still a matter for debate[20], but it is a fact that scientists played an important role in disseminating news of the Bikini hydrogen bomb experiment disaster of March 1, 1954, in which the crew of a Japanese fishing boat were injured by nuclear fallout. This led to a nationwide antinuclear movement.

Despite Japan being the only country which has fallen victim to the atomic bomb, knowledge of the harmful effects of nuclear weapons was not widespread because of the Occupation policy of suppressing any protests against the use of the bomb. Even the leading leftist scientist, Mituo Taketani, commented right after the war that the atomic bomb was the heavenly punishment for Japanese militarism provided by the world comradeship of scientists. The Bikini case constituted the beginning of Japan's grassroots opposition against nuclear policy.

On the other hand, as the impoverished standard of research and living standards progressively improved, the typical member of the scientific community turned out to be apathetic towards political action. In the Japan Science Council, the *Minka* school still continued to be active, especially in the fields of history and economics, whereas the Marxist school has been influential in postwar Japanese academia.

Though *Minka* delegates by no means held the majority of votes, their actions were quite visible, as most of their energy was concentrated in issuing pacifist political manifestos. Apolitical members of the Council could do nothing but agree. Most nonpolitical, non-ideological delegates, especially in the applied science fields, represented the interests of their own profession in a more or less technocratic way. The main reason why these seemingly incompatible groups of people could co-exist within the JSC is because their areas of interest did not overlap and therefore were not in direct conflict with each other. In the meantime, most of the power of the JSC was eroded by the bureaucratic agencies, such as in the case of the Ministry of Education's increase of its control over research fund allocations and the STA's promotion of large national projects. Consequently, members of the scientific community, particularly those from the younger generation, lost interest in the JSC, as evidenced by the decline in the voting rate for the election of new delegates. Its peak period was in 1950 and 1953 when more than 90% of the total scientific community voted. This decreased to less than 66% in 1971. Furthermore, the average age of delegates increased due to the lack of interest on the part of the younger generation.

I. Changing from Labour-intensive to Capital-intensive Science

In the mind of the Japanese scientist, the Japanese postwar constitution prohibits rearmament and hence only Japanese scientists could solely devote themselves to the peaceful use of atomic power, without being distracted by its military application. Thus, most scientists in the 50s welcomed the full-scale development of nuclear research. Around the same time, influenced by the Bikini incident in 1954, local residents living close to a proposed nuclear research institute protested bitterly against it and *Chidanken* supported the residents under the banner of 'national science.'

II. Domination of Technocracy over Democracy in Science (late 50s to early 60s)

Around 1955, the productive capacity of Japanese industry returned to prewar levels. In the late 50s, the job market for scientists flourished. Internationally, the late 50s was the period in which most industrial countries vigorously promoted science and technology, especially following the post-Sputnik science and technology boom of the USA. Japan now attempted to catch up with the rest of the world. Japanese technocrats, represented by Seiji Kaya, began to gain power in the late 50s and they were to find their happiest days in the early 60s when the government supported them in full swing to create a science and technology boom for economic growth.

1. Democracy Alienated – the Mechanism of the *Minka* School

There is a fundamental problem in the organization of scientists that the *Minka* group raised. In the period immediately after the War, 'democracy' was quite often used rhetorically to mean something good, something new against the old feudalistic regime which prevailed in prewar Japan. Democracy also meant a 'bottomup' institution as exemplified by unionistic ideals.

In order to achieve its political objectives, the *Minka* group adopted the strategy of securing a majority of the votes in the parliament of scholars, the JSC. Democracy is institutionalized in the political institution of majority rule. Whether unionistic or political, the principle of solidarity was maintained to enable a like-minded group to achieve the greatest impact within this institution.

Applying the principle of solidarity within the academic associations where it had members, *Minka* was able to control the election of the delegates from these associations. Since apolitical members were not organized, the well-organized *Minka* always earned more seats than their real strength deserved, in the election of the delegates of the JSC. As the next step, the elected few of the *Minka* group politicized the organization of the JSC by applying the same tactics and strategy of their election campaign to the internal politics of the JSC.

The apolitical delegates were dismayed, but they did not have the zest to organize a group to oppose and counteract the dominance of *Minka*. As far as a democratic political institution is concerned, the strategy of the *Minka* group was completely legitimate. The average scientists, however, did not believe that their organization was primarily political. They simply quit the organization when it did not represent their common professional interest, such as the securing of research funds.

Many of these average delegates to JSC tended to think of

themselves as selected elites; they represented their profession not because of their political stance but because of their high academic performance, which has nothing to do with democratic procedures. Quite often, their elitist stance conflicted with the interest of the majority. But in the community of academic science, the prestige of high performers is acknowledged even by deep devotees of academic democracy. Thus, alienation of the apolitical scientific community from the political and democratic *Minka* group became more apparent over the course of time.

In the election mechanism as such, the technocrats in the government and the ruling political party entertained misgivings that the JSC was dominated by the *Minka* group and ultimately by the JCP; the technocrats challenged the very constitution of the JSC in the early 80s, and in spite of stout resistance by the *Minka* group, the government succeeded in forcing a change in the selection method of delegates from one of direct vote by the scientific community to the recommendation of scientific societies. Consequently, the *Minka* group lost their position and were nearly eradicated from the JSC delegation.[21] It was a symbolic indication of the demise of postwar democracy.

Even now there still exists a kind of taboo in academia on disclosing who is a *Minka* member (later, it is generally nicknamed *Minsei*, an abbreviation of *Minshu Seinen Domei*, League of Democratic Youth, an organization of young Communist party members). The Japan Communist Party, still retaining some of its underground flavour from the prewar period, does not disclose its membership directory, which invites the suspicion of non-members. The Party, rather than promoting democracy within the Party, adheres to a top-down command system which discourages freedom of speech and induces the loss of the more original and creative members.

In the larger political context, the *Minka* were responsible for establishing the orthodoxy of Marx-Leninism by themselves in the midst of Sino-Soviet strife from the late 50s on. Consequently, the *Minsei* group tended to be intellectually conservative and apologetic in maintaining the orthodoxy of postwar democracy. Some timelag was necessary before they could revise their principles in response to the changing circumstances. They inevitably lost the intellectual cutting edge.

From my observations during the 70s and 80s, those faithful followers of the *Minsei* are sober and serious students without much originality, while those new-left critics are flamboyant free-thinkers. In Japanese academia, earnest, stereotyped followers of leftist ideology are often mistakenly thought of as having a *Minsei* affiliation.

III. Democracy vs. Technocracy in Science (from late 60s on)

The late 60s marked a watershed. The foregoing technocratic phase was eclipsed and became the subject of criticism by the resurgent grassroots democracy movement occurring in most Western industrialized countries. No historian has given a clear-cut explanation of this phenomenon which occurred simultaneously in many corners of the world. The anti-Vietnam war campaign, anti-pollution and environment movements, university unrest and reform, the influence of the Chinese Cultural Revolution, the intellectual current of anti-science, counter-culture and Hippyism, the change of priority in national science policy – all these factors combined to create a turning-point in the relationship between technocracy, science and democracy. Japan, though located in the eastern extreme of contemporary civilization, was no exception.

A. Creation of *Nihon Kagakusha Kaigi* (Association of Japanese Scientists)

Along the line descending from postwar *Minka* activity, the *Nihon Kagakusha Kaigi* (hereafter abbreviated to NKK) was created in July 1965, with the primary aim of protesting against the Vietnam War and particularly against the use of biological and chemical weapons. The JCP wanted to have an association of scientists and intellectuals similar to the old *Minka* to represent the concerns of the Party, independent of the Soviet Union and China. Those scientists who were called on to organize this group were concerned that too much politicization might invite resistance by the scientific community and intentionally avoided giving the impression that *Minka* had been reborn. They hoped to unite Japanese scientists, regardless of differences in ideology and research methodology. In view of past conflicts between different branches, such as those between the physics and geology communities, they wisely refrained from creating specialized branches. Local juntas became the basis for organizing their major activities. These provided a setting where problems of science and society could be discussed at the level of local scientific communities rather than in a centralized hierarchy.

B. Scientist Attitudes Classified

In the postwar period, Japanese academic scientists have been fortunate enough not to have been involved in military research. At the time of the Vietnam War, most Japanese scientists and engineers were not involved in the war effort. As intellectuals, however, scientists had the opportunity to express their own views. Scientists can be broadly classified into the following three categories, depending upon their way of thinking and behaviour on the matter of science in a social context.[22]

1. Academics

Japanese classical academic scientists believe that their trade is 'neutral', independent of any social context. They feel some professional obligation to contribute to the scientific community they belong to, but not necessarily directly to society at large. Academic science is purely an activity based upon individual performance; hence the scientists are inclined to view social problems from their own vicwpoint.

They would not be particularly fascinated by military research. When the problem of military funding is raised, they try to discriminate between 'political' and 'scientific' issues, and attempt to concentrate on the latter. As long as they can avoid being involved with 'dirty' research funds, as long as they can keep their own hands clean, their conscience remain untroubled and they feel safe and secure.

2. Unionists

The unionist type of scientists are those who support the JSC[23] and NKK. They do not view the issue purely as an individual matter. Often they feel that there is nothing wrong in the promotion of science but that its utilization is quite distorted and abused because of poor social institutions. This is their basic view of science, the origins of which can be traced back to the Minka school.

As stated earlier, their paradigm is J.D.Bernal's *Social Function of Science*, which gives a strategy by which to 'modernize' scientific communities, organize research, reform society by means of science and eventually bring forth the final great achievement of the world of communism. The social mode of science, in terms of promotion and utilization, has changed considerably since the time of the first printing of Bernal's book in 1939, *Minka's* heyday around the 1950s, and the time of NKK in the late 60s.

While academics tends to limit their energies to the promotion mode and reject too much involvement in utilization, unionists both promote and utilize and consider the two aspects from a viewpoint which opposes and confronts capitalistic society. In their scheme, the promotion of science in a scientific community guides and directs the utilization of science in a capitalistic society. By so doing it strives for social reform or revolution. In the words of 'unionists' such as Mituo Taketani, only those who promote science are qualified to direct the utilization of scientific achievements; when science is properly utilized in a society, it is seen as leading to an increase in production capacity, which in turn invites inevitable social change. Conversely, when society changes, science is properly utilized. Right after the War, the Japanese were so impoverished that the utilization mechanism of science was diminished and powerless. Democratic scientists thought that an initiative in reforming the utilization of science would lead to the recovery and eventual reform of the whole society. This unwarranted ideal was partly shared by technocrats, who also aimed for recovery through science by appealing to the establishment rather than to citizens.

During the two decades between *Minka* and NKK, Japan experienced high economic growth without any social revolution. The utilization mechanism of science, which can mainly be seen in the big corporations, has grown in strength and subordinated the promotion mechanism of science, by enlarging the capacity and might of private industrial research.

From the viewpoint of *Minka*, the promotion of scientific research is in nearly all cases legitimate. *Minka* scientists accuse government and capitalistic society of negligence and abuse of science. On the other hand, scientists and the scientific community are encouraged to promote their research free of the burden of social responsibility.

For *Minka*, as well as NKK, the value of science is to be defended by all means. Since the mid-70s, the JCP gave up Marx-Leninism and replaced it with 'scientific socialism' in Engelsian fashion. Hence, communist members of NKK are in a position to defend science as Party directives.

Like a trade union in closed-shop style, such scientist-organizations defend the standpoint of scientists from the criticism of society, including that of citizen level. In a Bernalian unionist strategy, scientists should organize their association of scientific workers domestically and join the world federation of scientists internationally. It is thought that by doing so, scientists with the internal solidarity of their own organization can demand more from the capitalistic establishment in a class struggle. The JSC was a democratic body elected from and by all of the Japanese scientific community. It was natural that it should take a union-like stance.

In order to maintain solidarity and the authority of the scientific community, internal criticism and internal reform by members is often discouraged. When the interests of the scientific community and grassroots society conflict with each other, it may invite the isolation of the latter community from the democracy of the larger society. Unionists of postwar democracy are not well prepared to resolve conflicts. If deprived of ideological aspiration, their organization would tend to be a mere professional association similar to a medical association.

3. Populist

At the leftist extreme, there is another group of 'populists' who have radically different ideas from the unionists. The JCP in the early 50s promoted a populist paradigm and accordingly *Chidanken* adopted a 'National Science' enlightenment strategy with a top-down hierarchy of command. While *Chidanken* failed to find grassroots followers, the new populist left, who arose in opposition to the old left of the 1970s, are much closer to less ideological citizen movements such as the anti-nuclear reactors campaign [24]

The populists do not recognize the neutrality and objectivity which academics attach to their science. They claim instead that the promotion of science is nothing but a question of ideological and political choice. Unlike the unionists who distinguish between good promotion and bad utilization of science, populists define promotion as a subsystem of utilization, as can be seen in the case of an industrial research laboratory of a big corporation.

Scientists consider themselves as guardians of objective truth. To a populist, on the other hand, scientists are no more than rank and file members in the existing establishment. While unionists saw conflict between the promotion and utilization of science, populists found conflict occurring between the establishment, in which technocrats and scientists are all incorporated, and the citizenry. Citizenry can be defined, in this context, as those who have no vested interest in the promotion and utilization of science.

These populists are by no means subscribers of anti-science or occultism, as they have sometimes been accused. Rather, their intentions are to take incorporated science back to the citizen level with slogans such as 'science for the people'. In the populist interpretation, 'science for democracy' in its true sense is a slogan of their own, while the unionists' approach tends to be closer to 'science for technocrats.' In this sense, populists are properly called antiincorporated scientists rather than anti-scientists; still they have found difficulty in finding their niches in the scientific community, as they try to appeal directly to the citizen level rather than to their fellow members of the scientific community; this action is often considered to be a violation of the rule of the scientific community as a closed union.

Consequently, populists turned out to be often critical of the scientific community. In reality, we find science journalists are more likely to be populists than practising scientists, since journalists write for and appeal to the citizens while scientists by profession write only for the research community they belong to; there is a certain invisible community moral code and self-restraint that works on the members of the scientific community. Other populists come from the group of sociologists, historians, philosophers of science, or, in general terms, STS (Science, Technology and Society) people, who belong to academia but who do not have to be loyal to the scientific community. I give a generic name to these people: 'para-scientists'.

In the history of ideas of social revolution, science was considered by Engels and Lenin to be at the vanguard, but in recent history Mao Tse-tung and Castro have not talked a great deal about science.[25]

In the meantime, science has passed its role of enlightenment and to become incorporated into the establishment of advanced countries. Science is contributing to the ever growing gap between developed and developing countries, strengthening the oppression of the have-nots by the haves. While Bernalian unionists are still optimistic about the role of aid from advanced countries in solving third world problems, populists seek self-reliance.

In general terms, we may say that old left unionists still depend on the old strategy that Bernal and *Minka* designed. This was perhaps because they were slow in responding to the new problems in the relationship between science and society and took the apologetic stance of conserving the old way of scientism. On the other hand, new left populists are keener than ever to find out new social problems which have been caused by science, but they have yet to establish a consistent policy and strategy for uniting the populist front.

4. Three Schools Compared

Although there are no actual statistics available for the members of each school, the majority of the scientific community belongs to the realm of academics, while unionists can be counted a power of ten lower, and populists another power lower, even with the inclusion of sympathizers along with the explicit proponents. These three groups are not unique to Japan as parallels can be found in most advanced capitalist countries. Perhaps unionists may be a unique aspect of Japan and a legacy of postwar democracy, though we may find some similarity with the British Association of Scientific Workers, which now unionizes laboratory technicians and secretaries.

In the following list, we shall contrast the differences of the three groups rather simplistically for the convenience of comparison.

	academics.	unionists	populists
position of science	neutral	leader of society	subordinate to establishment
promotion of science	professional duty	absolutely good	to be assessed by people
utilizers of science	Scientific community	technocracy	people
political stance	apolitical	progressive	radical

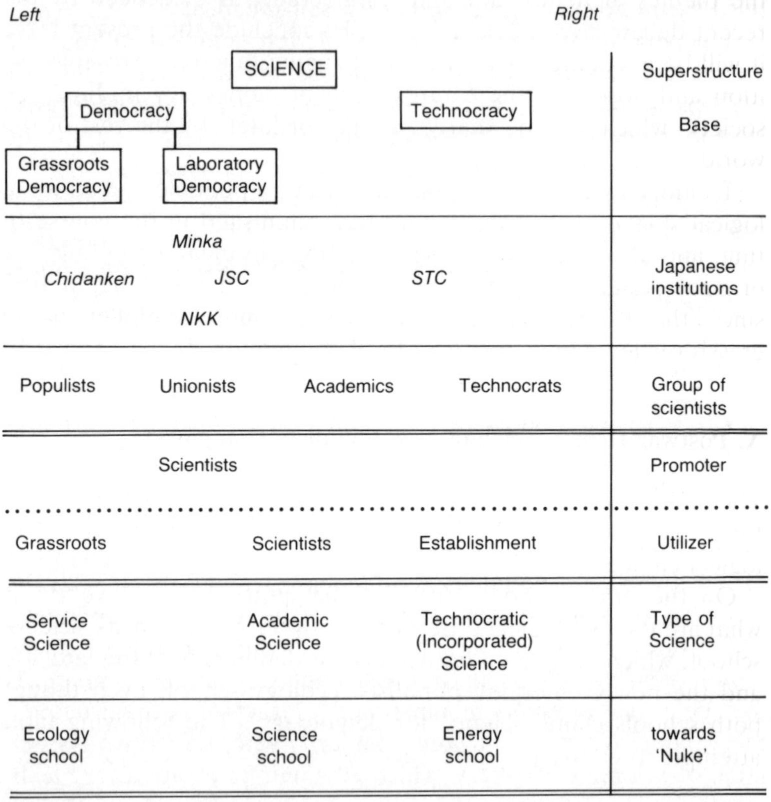

Figure 2 Chart explaining the structure of the Japanese scientific community

IV. Conclusion: The Place of Democracy in Japan

We have so far surveyed how science was placed between democracy and technocracy, in the first period of postwar democracy in the late 40s to the early 50s, in the second period when technocracy gained power over democracy in the late 50s to the early 60s, and in the third period when grassroots democracy challenged technocracy in the period from the late 60s on. We are now entering into the period of articulation and confrontation between the themes of democracy and technocracy, as evidenced by the recent debate over nuclear issues. To conclude the present topic it will be adequate to discuss the changing function, present position and hopefully the future vision of democracy in Japanese society, which may be shared, sooner or later, by the rest of the world.

'Democracy', which was a catchword or a political and an ideological slogan in postwar Japan, has diminished in the course of time and also undergone a subtle change in meaning. While the original American version of democracy is diminishing, it has, since the Chernobyl incident, become more endogenous in grassroots level movements in local communities.

A. Postwar Democracy and New Populism Compared

The difference between democracy and technocracy is crystal clear for everybody. Which side science belongs to is a matter of ideological choice.

On the other hand, it may not be quite clear to everybody what are the ideological differences between the unionists' science school, which originated in the postwar democracy of the late 40s, and the newly emerging populists' ecology school, even though both schools claim 'science for democracy'. The following table attempts to clarify the difference.[26]

(S: science, T: technology)

postwar democracy	critical populist
good science, bad social system	science and social system indistinguishable
scientific advancement leads to change	scientific advancement to maintain the social status quo to serve the establishment

postwar democracy	critical populist
universality & neutrality of S & T	local and ideology-bound ST
anti-feudalism, modernity	recovery of communality
strengthening of autonomy of S	break of autonomous S
inseparability of scientific rationality & democracy	inseparability of scientific rationality & technocracy
enlightenment and dissemination of information of scientists on autonomous standpoint of scientists	utilization of specialized science by the iniative of citizenry
improvement of rights and status researchers	self-negation & self-reform of researchers
establishment of social status as a professional	critique of anti-social character of scientists as professionals
efficiency of intellectual production	relevance to life rather than efficiency
science for people	science by people
increase of professional manpower	rejection of division of labour
social responsibility of scientists as elites	social assessment of ST
social reform by the initiative of scientists	reform of science on the initiative of people
promotion of peaceful application of nuclear power	mankind does not coexist with nukes
laboratory democracy (within scientific community)	
democratization of scientific activity (with participation and intervention of citizenry)	
critique of irrationalism & emotionalism	criticism of scientism as an ideology
autonomy & priority of S over politics	mode of S reflects mode of politics
strengthening ST, economic recovery through industrialization	preservation of nature, emphasis of bio-environmental S
respect for Western science	sympathy toward third world
imported or national T	appropriate T

We could continue this table almost endlessly. Having listed the differences, I am myself surprised at how the gulf between the two camps has widened in the course of some forty years. This course of history is irreversible and it is unthinkable that the two schools will unite again, as the socio-political environment in which science is placed has undergone a major transformation, from democracy to technocracy in many parts of the world, including Japan.

From the viewpoint of democracy science has also been transformed from a liberator to oppressor in its social function. The social function of science, whether in the Bernalian sense or not, directly reflects the point-of-view of the utilizers. For those people who have devoted themselves to science for its own sake and even in the classical Bernalian sense, the social function of science was formerly confined to that of academic science, in which the utilizers are themselves scientists. The postwar development of science gives evidence of constant change from academic to industrialized or technocratic (incorporated) science, in which traditional academic rules do not hold. Instead, non-academic utilizers dominate the decision-making of research objectives and the major line of development. These non-academic utilizers form a technocratic regime, assisted and advised by elitist spokesmen of the scientific community.

On the other hand, the promotion mechanism, in which scientists and engineers are directly concerned, has turned out to be a mere sub-system of the evergrowing utilization mechanism of states and enterprises. At this time, there have emerged sharp differences in attitudes towards science amongst scientists. Academics and even postwar democratic scientists pursue traditional lines of scientific development, primarily aimed at assessment by the non-academic technocratic establishment, while populists try hard to find an assessment and utilization mechanism for their science at the grassroots.

The present utilization mechanism by technocracy is so monstrous that it has swallowed the promotion mechanism of postwar democratic scientists, though they may not yet be conscious or concerned about this development; only a handful of populists have emerged to oppose the technocratic trend with their Don Xihotean spearhead.

Notes

1 See the work of L. Hoddeson and S. Traweek.
2 Dorothy's diary, Nov. 24, 1945, from Joseph and Dorothy Needham, *Science Outpost Papers of the Sino-British science cooperation office, 1942–1946*, (London 1948).
3 GHQ, ESSST sources (preserved with index at the Diet Library, Tokyo).

4 Charles Weiner's interview with Harry Kelly, cited with permission of the MIT Archives.

5 Shigeru Nakayama, 'Harry Kelly' *Kagaku Asahi* September 1987.

6 Interview with H. Kelly.

7 Shigeru Nakayama, *Academic and scientific traditions in the West, China and Japan* (Tokyo University Press, 1984), chapter 1.

8 E. Reischauer's lecture at Harvard University, 1955.

9 Nihon Kyosanto Kagaku Gijutsubu (Science and Technology Section, Japanese Communist Party),'Nihon no kagaku gijutsu no kekkan to kyosan shugisha no ninmu' ('The defects of Japanese science and technology and the role of communists'), *Zenei*, no.10, 11 (November, 1946).

10 With the problematique of 'science and democracy in conflict' a Euro-Japanese dialogue of parliamentarians and scientists was held in Tokyo, Japan in June 1985 with the sponsorship of the Council of Europe, though Japanese participants did not seem to comprehend the issue well enough.

The proceedings were published as *Europe-Japan, Future in science, technology and democracy* (Council of Europe, 1986) edited by Vincent J. McBrierty. On p.9, Sir Hermann Bondi states that 'A vital assumption in both science and democracy is that opinions can be altered by discussion, by argument and (above all) by empirical evidence. Indeed, discussion and its necessary precursor, disagreement, are essential for scientific no less than for political progress.' Here, he means academic science, not technocratic science.

11 Shoichi Sakata, *Kagaku to shakai* (Science and society, Iwanami, 1972) p.3 ff.

12 Shigeru Nakayama, 'Grassroots geology' in S. Nakayama et.al.(eds.), *Science and society in modern Japan*, (MIT and Tokyo University Press, 1974).

13 L. Hoddeson, 'Establishing KEK in Japan and Fermilab in the US: Internationalism, Nationalism and High Energy Accelerators', *Social studies of science*, vol.134 (1983).

14 Shigeru Nakayama 'The American Occupation and the Science Council of Japan' in Everett Mendelsohn (ed.), *Transformation and Tradition in the Sciences* (Cambridge University Press, 1984) pp.353–369. The official account of the activity of the Science and Technology Division, ESS was written by Dr. Putnum in a 'Non-military History', the most part of which is devoted to the establishment of the Japan Science Council. It represents a self-evaluation of the Division's work.

15 Shigeru Nakayama, 'The American occupation and the Science Council of Japan', in Everett Mendelsohn ed., *Transformation and Tradition in the Sciences* (Cambridge Univ. Press, 1984) p.361.

16 Kaya also achieved the nickname of 'a stamina democrat', which implies that while he chairs a committee, he is tolerant up to the last minutes when everybody gets tired of debating, and then, he finally decides. His own thought was, however, much closer to that of technocrats, whose trans-political nature obscures explicit differences.

17 Kagaku Shinbun (*Science News*).

18 Shigeru Nakayama 'American Occupation and Japan Science Council' in Everett Mendelsohn (ed.), *Tradition and Transformation in Science*, (Cambridge University Press, 1974).

19 Haruhiko Hamaguchi, 'Minshushugi kagakusha kyokai to sengo junen' ('*Minka* and its early ten years'), *Shakai kagaku tokyu*, vol.25, no.1 (1976) pp.91–112.

20 Chikayoshi Kamatani and Tetsuro Nakaoka, 'Sengo kagaku gijutsusha undo no sokatsu' ('The recapitulation of movements of postwar scientists and engineers'), *Gendai no riron*, vol.7, no.9 (1970) pp.36–55.

21 Shigeru Nakayama, 'Gakujutsu kaigi wa ima' ('The present status of the JSC'), *Kagaku*, vol.59, no.1, (1988) pp.45–49.

22 Shigeru Nakayama, *Kagaku to shakai no gendaishi* (A contemporary history of science and society, Iwanami, 1981) pp.34–40.

23 Erik Baark and Andrew Jamison (eds.) *Technological development in China, India and Japan* (Macmillan, 1986) p.239.

24 Jinzaburo Takagi, Jun Ui and Atuhiro Shibatani are counted as leaders and ideologues of this group.

25 I. Bernard Cohen, personal communication.

26 Shigeru Nakayama, *Kagaku to shakai no gendaishi* (Iwanami 1981) p.231. In this table, We have put some of the characteristics of the old academics into postwar democracy, while the critical stance of the populist is emphasized for the sake of comparison.

3 Changing Models of Japanese Universities

One of the most effective reforms during the Occupation years was in education, where there was a transition from the old Japanese system to a new American-inspired model. In this chapter, we shall evaluate its effect on the training of scientists and engineers. From the late 1950s, the industrial sector demanded an increase in scientific manpower. We shall trace how Japan coped with such demands.

The Model of Japanese Universities Changes during the Occupation

Since Japan was neither colonized nor dominated by a single Western power before World War II, there was no colonial-type authority to enforce adoption of a particular existing foreign model until the American occupation period of 1945–1952.

Western models for Japanese universities are a favourite topic of discussion amongst Japanese academics and it is generally believed that prewar Japanese public universities are modelled after nineteenth-century German universities. If we examine this belief closely, we will find it not exactly correct, as the strong German influence came a little later in the 1880s, except in the case of medical schools, while American, British and French models were influential in 1870s in forming the basic institutional structure, such as the rigid residence requirement. The truth is that, in the 1870s, Japanese bureaucrats investigated and 'window-shopped' for various Western models, mixed them together and fitted them into the Japanese mould of bureaucracy.

While the Western models of universities have a tradition older than the bureaucratic system of modern states, modern Japanese universities of Western style are purely a product of the Meiji government. They were developed as an integral part of Japanese bureaucracy. Thus, the true model of a Japanese university is not that of any particular university of a particular Western country but of bureaucracy itself.[1]

In the postwar period, Japan adopted the American model of education. In contrast to most third-world countries, where modelling after the 'master' country came first, followed by the autonomous choice of model in the postwar period, the Japanese experience was exactly the opposite and therefore quite unique. While the former pattern is the transition from 'dependency to autonomy', the Japanese experience was one which was diametrically opposed.

Four factors of the American educational model would be analysed: 'democracy', 'egalitarianism', 'decentralization' and 'lay control.' Before the advent of postwar American influence, the Japanese had already established their own bureaucratic model. The way the American model and concepts have infiltrated this bureaucratic model is one of the major topics of the present chapter.

'Democracy' and 'Egalitarianism' were translated in a higher education context into the transition from elite to mass education. During the Occupation period, the Japanese bureaucracy reluctantly accepted this trend that to some extent prepared and pre-emptively met the later expansion of demand for higher education. It also resulted in creating a 'cheap universities' tradition. Bureaucracy insists on planrationalisation and an entrance quota rather than meeting market demand. Further expansion in the high growth period was left to the private sector.

'Decentralization' was dismissed, on the grounds that local governments lacked the appropriate resources to handle matters of higher education. 'Lay control' was rejected in an effort to avoid capitalistic and militaristic influence on campuses.

Under the circumstances mentioned above, I shall explore how Japanese academic science and technology has evolved.

From Elite to Mass Universities

By the mid-twentieth century, Japan had firmly established its own institutional paradigm of higher education and hence, its replacement by a new American model could not be achieved without the application of extraordinary pressure by the Occupation Forces. Educational reform differed in certain respects during the postwar Occupations of Germany and Japan. An old elite Japanese who had sympathy towards European education said that while Germany refused to accept the American model of postwar reform, the cowardly Japanese accepted it without any sign of resistance. The fact is that, even though the Occupation Forces consisted of representatives of several Allied powers, the Occupation of Japan was administered almost exclusively by the US, in contrast to the Occupation of Germany, where French delegates were opposed to the exclusive introduction of the American model. Thus, the postwar reform was not an indigenous reform but one enforced by the Civil Information and Education Section (CIE) of the American Occupation and their advisor, the American Educational Mission, for whom it was simply inconceivable to propose any model other than the American one, with which they were the most familiar.

Americans considered the prewar Japanese educational system to be too centralized. They seemed to entertain the idea of disbanding the Ministry of Education, the hub of the power structure. But the American Occupation decided to carry out its reforms through existing Japanese offices in order to maintain the forms of indirect governance. They advised the Japanese higher education reformers to create the University Accreditation Association in 1947, which was instrumental in the initial step towards higher education reform in the American mode.[2] The relationship between the American CIE and the Japanese Ministry of Education was an ambivalent one.

The Association was active in setting the goal of and guiding postwar university reform in its first decade but actual inspection and evaluation of new-system universities were carried out by the University Chartering Committee (later renamed Council) organized by the Ministry of Education. Still later, its function was transferred to the Ministry of Education, which enacted the Standards for the University Charter in 1956.[3]

In history, human resources matters can never be handled as

planned. The Americans gave up their original intention and withdrew their proposals, when Japanese opposition became too strong. They intended to abolish the university chair system but failed. Their proposal for the introduction of the American-style Board of Trustees as the governing body of public universities was ruled out, in view of the opposition of Japanese authorities who insisted upon the autonomy of the individual universities.[4]

Their fundamental purpose of reform was the replacement of ultra-nationalistic and militaristic ideology with democratic thought. As discussed in Chapter 2, the democratization trend was, in principle, accepted by most university scientists, who were never ultra-nationalists, nor militarists. In implementing the American model, however, economic realities caused difficulties.

Expansion of Higher Education during the Occupation

In prewar Japan, graduates of higher education comprised fewer than 7 percent of the age cohort and hence their education was naturally elitist. The prestigious Imperial and other national universities, enrolling less than 1 percent of the age cohort, were the site of most prewar academic science, where the nineteenth-century German Humboldtian ideal of unification of research and teaching was maintained.

In 1946, the first United States Education Mission to Japan advocated a singletrack, 6-3-3-4 school-year system from elementary to higher education, instead of the prewar 6-5-3-3 system. They also advocated that the Japanese expand their higher education sector to the extent that each prefecture (46 in number) would have its own university, apparently modelled after state universities in the USA. While prewar universities were located in major cities, the egalitarian principle of equal opportunity for higher education was promoted with decentralization of educational facilities to include the new provincial middle sized cities.

Incorporating this strategy into its planning, a significant expansion of higher education was proposed by the Ministry of Education, in spite of the economic hardship that prevailed after Japan's defeat in the war. It generated numerous complaints and a great deal of confusion in Japanese academia.

In actual practice, the last two steps (3-3 years) of the old education system were now to be converted into the last step (4 year universities) in the new system. This move certainly created a great opportunity for the old professional schools (the third step of 3 years training in the old system) and normal schools to raise their status to the university level. They took advantage of the opportunity and most were successful in obtaining accreditation as universities.

Private sector education was still more aggressive, flexible and adjustable. Twelve private universities, mainly in the Kansai area, announced that they were ready to start as new-system universities in April 1948, prior to the response in the following year by most public-sector schools.

Consequently, the Japanese higher education sector was expanded, within a few years, to one that exceeded ten times that of the prewar sector, in terms of the number of universities, students and teaching staff.

The prewar multiplication of Imperial universities, which took place nearly every decade, had always brought new perspectives and new meaning into the Japanese academic sector, such as the introduction of new disciplines and the enrolment of women students,[5] forbidden in older universities.[6] In contrast, no new-system university during the postwar expansion offered new perspectives and meaning (perhaps with the sole exception of the International Christian University, which was a new institution run in a purely American way) but instead assimilated themselves to the older universities in order to be accredited and conferred with a university charter. They could by no means breathe fresh air into the existing scientific community as the newlycreated Imperial universities had done earlier. If we consider the reforms *in toto*, however, a new model was certainly created; that is cheap, democratic and low-key mass higher education, replacing the old system which had been of elites, by elites and for elites.

Cheap Universities

Since the inception of the modern Japanese university system in the late nineteenth century, the demand for higher education from the citizen sector has been constantly rising, often overtaking the planning and preparation of facilities by the public sector. It was, however, nearly impossible to plan such a rapid expansion at the time when old facilities had not yet recovered from the destruction of wartime air-raids. Most of the energy of faculties and students was spent looking after their own subsistence. These new-system universities were naturally very poorly equipped in terms of both teaching staff and facilities.

The government budget, throughout the postwar period, simply could not cope with the expansion of higher education, a situation which inevitably invited the impoverishment of universities and university research. The Ministry of Education commented in 1964 on this state of affairs in the following way: 'Since the installations and equipment were in this condition, it was inevitable that the operating expenses of the universities were reduced to a bare minimum, and since most of the available revenue had to be spent on salaries of faculty members, special allowances, maintenance expenses and administration, the sums used to educate students and to carry on research activities were really nominal.'[7]

The traditional academic elites, whose concept of higher education was still elite-centred, criticized these new universities as being second-rate and expressed their grave concern at the deterioration of the quality of university research and teaching. Confusion and misunderstanding of old elites, who graduated from the prewar Imperial Universities, in the new system lasted for more than ten years.

The extraordinary expansion of higher education in a devastating economic context resulted in a major change in the traditional function of universities. University faculties wished to maintain the high academic standing that was expected from elite students in the old system. Those elites were qualified to study by themselves in the old German way of academic freedom without being inducted in the American educational style of spoonfeeding. Whereas such an elite student culture could no longer be maintained in the new mass universities, faculties did not develop the American way of teaching, leaving non-elite students untrained and uneducated. Students consequently were at a loss as to what they should aim for. University teachers who had little interest in training students too easily gave credits and diplomas, compared with the practice of American universities. Consequently, universities were transformed into massproduction-based diploma mills, cheap in quality and cheap in expense.

This notion of universities, cheap in facilities as well as in the quality of education, which was rooted in postwar economic misery, lingered even into later days of affluence. This notion certainly contributed to the impoverishment of university research in comparison with private industrial research laboratories. If we compare the laboratory expenditure (per chair) of experimental science at major prewar governmental universities with the consumer price index, we find that university research in terms of budget has suffered for a long period, unable to catch up with postwar inflation (Fig. 3)[8], only equalling the prewar standard in 1965 after which there was no significant increase and at times even decreasing.[9] Non-experimental courses at major universities have not surpassed the prewar standard. The research budget of postwar new-system universities is still far inferior to that of the prewar universities.

Selection Rather than Education

Non-elite student bodies assembled in new universities spend their four-years enjoying freedom and in repose, emancipated from preparation for arduous college entrance examinations. As Japanese universities are hierarchically arranged according to the quality of their faculty, the university which one enters may very well determine one's future career and hence all primary and secondary education is geared to this end. Perhaps the most important function of Japanese universities is selection and labelling of student quality at the time of the entrance examination rather than during or after four years of education. Personnel officers of Japanese corporations, who are major employers of college graduates, only evaluate the selection function and pay little attention to student achievement during their four years of university education.

Foreign observers praise pretertiary education as superb in contrast to the poor quality of university education, while internal criticism of 'examination hell' prevails in the mass media, attacking the 'inhuman formalism' of preparation for examination in secondary schools that kills the students' creativity and aspirations.

Japanese high school education was internationally praised in an international comparison of mathematical scores conducted by UNESCO, in which Japanese excellence was shown.[10] Japanese high school students are strong in subjects like mathematics, which require, more than any other subjects in school examinations, a

Figure 3 Laboratory budget

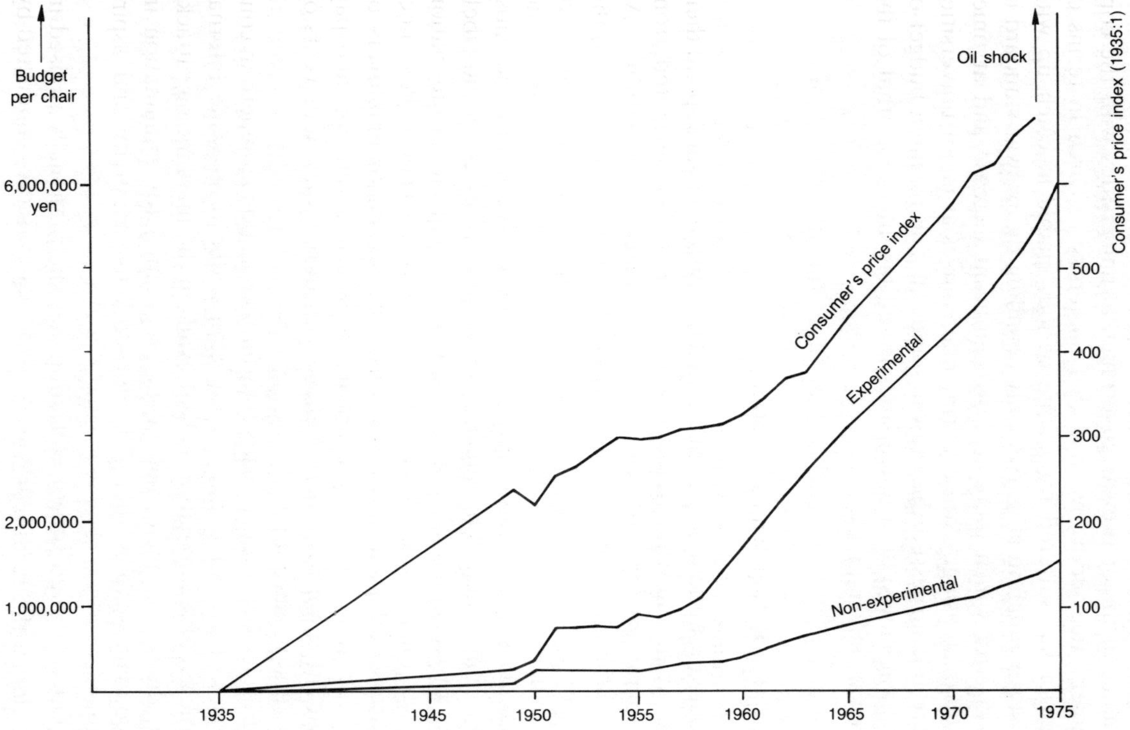

Sources: *Wagakuni no Koto Kyoiku* (Higher Education in Japan, 1964), p. 55. *Wagakuni no Gakujitsu* (Arts and Science in Japan, 1975), pp. 80–81.

certain amount of step-by-step preparation. School mathematics is merely an indication of testing studiousness in an unrealistic problem-solving game, rather than genuine creativity. High achievement may be correlated with a later career in the scientific professions but those Japanese scientists who are tested at school mathematics may be good in solving given problems in normal scientific activity rather than identifying and solving problems.

On the other hand, students can enjoy four years of non-regimented freedom in their young sensitive days, between the earlier examination hell and later disciplined life in 'Japan Incorporated', though four years is too long a vacation. Like everywhere else, however, Japanese science and technology students are generally kept much busier than students in humanities and social sciences. As new disciplines are added on top of the old required courses, their curriculum becomes overcrowded and prevents students from gaining adequate mastery of any particular field. Again, personnel officers of corporations boast that companies re-educate their recruits.

Egalitarianism – Toward the Homogenization of Scientists

Before World War II, different tracks of education were designated for those who terminated their training in the primary, secondary or professional school (junior college) levels, largely depending on the students' social class origin. Hence, there were three levels of engineering: secondary, professional school and university graduates. Even in the field of medicine, practitioners were divided into two classes: university graduates and professional school graduates, despite the fact that both received the same medical licence. This complicated plural track of education was changed to a single track under the egalitarian policy of the American advisors. In the same spirit, the status of private universities and women's colleges was enhanced. Normal schools were given college status.

In reality, however, in view of the short period of preparation and the lack of facilities, it was impossible for the new universities to obtain the resources necessary to rise to the standard that had existed under the old system. Thus, in spite of the creation of a new university system, the unequal pattern of resource distribution that characterized the old system remained essentially unaltered.[11] There are differences in the quality of students between the descendants of old elite universities and newsystem provincial or private universities.

In the meantime, however, the nature of universities has undergone major changes from the elite university to mass higher education. Hindsight allows us to see that these new institutions preemptively prepared the way for the dramatic expansion of higher education that took place in the postwar years all over the world. One of the causes of the student revolts of the late 1960's in the Western industrial countries was interpreted as being due to overcrowding in universities. As far as the scale of student admission is concerned, Japanese universities suffered less than their European counterparts because the expansion had started earlier. Among advanced nations, Japan was the only country which had significantly expanded its higher education during the 1950s in terms of enrolment growth rates. Rapid growth in the tertiary education sector in other nations occurred in the 1960s.[12]

To the ordinary people, universities are no longer ivory towers, inaccessible to the citizen sector, but have been instead transformed into 'around-the-corner' institutions. At the same time, the homogenization of higher education has created a public notion that higher education is cheap, that it can be afforded without great financial burden on the part of the public or by the individual.

Graduate Schools

The American educational model is a successful system to promote science, a fact which is nearly unanimously acknowledged by the American scientific community. When they argue as to the shortcomings of American scientific education, their criticism centres on secondary education but never upon graduate schools.

It is significant to note that a Japanese graduate school affiliated to the Imperial University had been established as early as 1886. It may well have been modelled on the American system, given that such a system of graduate schools existed nowhere else at the time.

Japanese graduate courses were not structurally well-developed throughout the prewar era. Doctoral degrees were conferred by the Faculty Board, without any institutional relationship with the graduate school. No coursework was designed for the graduate school. The latter tended to function merely as a place where one could sit out the period until one was able to find a job. It remained an elusive institution, a remnant of a policy failure.

After the war, the American Scientific Mission, rather than the Education Mission, strongly urged that graduate schools be reformed.[13] The CIE of the Occupation Forces had formulated the model of the American graduate school system and provided guidance through a newly formed agency, the Japanese Association of University Accreditation. There was little resistance on the part of existing Japanese universities, because the old graduate school did not have any structure to conflict with the newly introduced American one. New graduate schools started in 1950 at private universities, which had been always looking for a chance for expansion, and in 1953 at national universities delayed by administrative rigidity.

Process of Internalization

After the withdrawal of the Occupation Forces in 1952, the Ministry of Education took over and strengthened its control over higher education. However, after seven years of model change, it was utterly impossible to return to the old model that had existed before World War II. The process of internalization (in other words, domestication of the external American model) has followed.

The prototype of prewar Japanese universities was established in the late nineteenth century with all its unique characteristics, such as faculty-department structure of universities. The minor changes that evolved were not enough to meet the new demands of the mid-twentieth century. A radical reform was desperately needed, regardless of its financial source. The American model was not necessarily new, as it was also largely formulated in the nineteenth century, but the newer elements were taken up by the Japanese; notably, mass higher education. If the elitism of the Imperial University had still remained, it might appear to be grotesquely anachronistic now. Other elements which were incompatible with Japanese higher education culture were excluded from the reform process. Hence, decentralization was discouraged in the period of internalization that followed and lay control of public universities was totally dismissed.

In higher education, the academic decisionmaking power which resided in the Faculty Board of each faculty, organized on the basis of undergraduate teaching, remained intact. The efforts of reformers focused on the introduction of new models in general education and the graduate school, new areas where old faculties had no vested interest. Thus, graduate schools were given no autonomous power, remaining mere subsidiaries to the organizations for undergraduates. Graduate programs were not well funded and graduate fellowships and assistantships did not adequately cover living expenses. Most graduate students support themselves by training high school students for college entrance examinations.

At the beginning of the new-system graduate school of the University of Tokyo, a new division (humanities, social science, mathematics, physical, chemical and biological sciences) was introduced to promote interdisciplinary research in contrast to the old faculties of law, economics, literature, science, technology and agriculture. But because the major decision-making power still resided in the Faculty Board of the old faculties, the structure was subordinated to and gradually restructured to fit to the old faculties division.

Most faculty members had no experience of study in the USA and hence were not well aware of the strong points of American graduate programs. Hence, coursework, reading assignments, the grade system associated with fellowship programs and other aspects of the infrastructure of the American graduate school are even now hardly systematized. The Japanese graduate school is still run in a manner resembling an apprenticeship, rather than systematic modern teaching.[14] Furthermore, it may be partly a matter of culture that Japanese professors do not train their students strictly, never failing them, and have failed to create a highly competitive atmosphere, as is the case in the U.S., even after more than three decades of its existence.

However, in the fields of science and technology, where there is more internationalization, than in the social sciences and humanities, new graduate schools based on the American model are more readily accepted. The graduate schools will continue to expand, since they can meet the future demands of science and technology, as well as the desire of raising the status of faculties which control the degree-conferring authority.

Entrance quota

Since the new system is not the product of spontaneous growth but rather the importation of an external model, there still are remnants of the old rigid bureaucratic model. This can be seen in regulations such as the residence requirement of two years for a master's course plus an additional three years for the doctoral program prescribed by the Ministry of Education, irrespective of the field and regardless of the individual ability of candidates. An entrance quota is also imposed for each discipline. It has only been in the late 1980s that more flexible graduate courses, such as those in the USA, have been contemplated.

As the system does not adjust to 'market demand' but rather, is designed by the bureaucracy to coordinate and maintain the status quo of existing disciplines, such rigidities do not meet the demands of a continually changing research frontier. Nor does it satisfy the demands of industry or indeed the citizen sector. It has been effective in suppressing newly-forming disciplines and serves to protect academic areas of declining interest.

A notorious example can be found in the preservation of agricultural disciplines. In the ten-year plan prepared by the Science and Technology Council in 1960, there was the alarming prediction that, for the following ten years, there would be a considerable surplus of agricultural scientists in contrast to the shortage in basic science, technology and medicine.[15] During the high economic growth of the 1960s, students were no longer attracted (perhaps except for agricultural chemistry, which has a large industrial application) and the nation's agricultural population declined. However, thanks to the entrance quota of teachers and students, these declining disciplines have been well protected. Since the discipline was not attractive, fewer students chose to major in them and the entrance quota was often not filled. Despite this, academic positions were maintained so that those who majored in an unattractive discipline had a better chance in academia in the future. In this way, the reproductive cycle of old disciplines is repeated without concern for the external pressures of society. Only in the 1970s were ecologically minded students and, in the

1980s, biotechnology inspired students, attracted to the agricultural science disciplines.

Whilst undergraduate entrance quotas are filled, those of graduate schools are usually not, indicating that the latter's status is not as yet socially recognized. The quality of a student is judged in Japanese society by the prestige of the university that admits them as a result of the university entrance examination. Graduate study adds little prestige except for those intent on pursuing an academic career. Thus, the Japanese graduate school is far less developed than those of Western countries. The ratio of graduate students to undergraduates is much lower in Japan compared with major Western nations. This is one of the causes of the cheapness of Japanese science.

The Training of Industrial Scientists

The industrial sector assists in supporting university Master's programs by providing special fellowships with the understanding that after the program candidates are guaranteed and obliged to be employed in sponsoring industries. Company engineers are also sent to graduate schools to obtain Master's degrees. In industry, Master's degree holders in science and engineering are industrial scientists, who are mainly assigned work in R&D and design departments, while Bachelor's degree holders are sent to the production line. Job transfers between the sections are also frequent. In the eyes of the Japanese corporation, Doctorate holders (Doctor of Science or Doctor of Engineering, and Ph.D. equivalents) are not welcome in industry as they are too narrowly specialized and too academically oriented to be good industrial scientists, an exception being those who are appointed for specific research that the company has a need for. Hence a master's program has developed, particularly in engineering schools, to meet the demands of the private sector (half of all master's degrees are given to engineering students), whilst the doctor's program has not undergone such a process. Since the mid-1980's, entrance quotas for engineering Master's programs have consistently been filled.[16]

One small episode should be related. A statistical study has been carried out which examines academic productivity in terms of scientific articles according to institutions, using *Chemical*

Abstracts.[17] Surprisingly, comparing the world's universities, the Universities of Tokyo, Kyoto and Osaka, rank among the best ten of the world. This is explained by Japanese scientists in the following way: no matter what the quality, Master degree candidates in Japanese universities are encouraged to publish articles on research topics which have been allocated or suggested by supervisory professors. This contributes to a high output rate of normal science reports per graduate student, as each student produces at least one article, even though students intend to be industrial scientists.

A higher degree does not enable a new employee to enter a company on a substantially higher salary level than one without this degree. Two years of Master's degree work is considered as being equal to two years of experience at the company under the seniority promotion system.

All of the above factors contribute to the formation of the character of postwar Japanese industrial science which will be discussed in Chapter five.

The way in which companies evaluate academic degrees has not undergone great change during the period from the 1960s to 1980s, though in this time the needs of high technology have certainly resulted in the upgrading of the academic qualifications of private industrial scientists. The academic sector, especially the faculty of science and engineering schools, claims that the frontiers of research cannot be reached simply with undergraduate training. They have proposed to upgrade training for scientists and engineers to the Master's Degree level. Some industrial sectors believe that they can do inhouse training much better than universities can but they have to have some tie with educational institutions in order to recruit new scientific manpower.

Manpower Policy During High Ecomomic Growth

Contrary to common belief, the postwar Japanese government did not have a comprehensive science policy, as far as R&D policy is concerned. MITI has been a protector and adviser of domestic industries but not a primary performer of private industrial science. The Science and Technology Agency sought to promote big national projects of public science but remained a small office.

The Ministry of Education is a promoter of academic science but remains isolated from the industrial sector. What is missing here, most of all, is the coordination among various government offices in formulating an integral science policy. If the Japanese government had a policy at all, it was the scientific manpower (increase) policy in early 1960s of the Ministry of Eduction, which had monopolistic power over the nation's education. Its prominent policy was implemented in accordance with demands made by the industrial sector during the period of high economic growth.

Industrial Demands for Scientific Manpower

As Japanese industry recovered in the late 50s, demands were made by the industrial sector for a more scientifically trained workforce. Postwar technological innovation accelerated this demand even further. In response to the industrial demand, the Ministry of Education made the first systematic calculation of future demand in 1957.[18] That was the year of Sputnik when the worldwide science and technology boom started. Based on the above calculation, the Ministry of Education announced the plan to increase science and technology student enrolments by 8,000 on top of the 22,000 entrance quota in 1957 to make a total of 30,000[19]. The plan was implemented year by year with an increase of 2,000 and completed in a four-year cycle from entrance to graduation in 1960.

It is the Ministry of Education which controls and sets the entrance quota of student enrolments for each public and private university. The public sector did not have sufficient funds to meet the student increase plan in only national universities, which are under the direct control of the Ministry of Education. To accomplish its plan, the Ministry called upon private universities to cooperate by increasing the entrarnce quota of their science and engineering students.

In November 1959, Kansai Keidanren (Kansai Federation of Japanese Economic Organizations) proposed a type of university reform which emphasized the specialization of technology teaching. In December, the Education Committee of Nikkeiren (Japan Employers Federation) demanded the creation of technical universities. In October 1960, the Council for Science and Technology presented a ten-year plan of science and technology promotion, in which an immediate increase in science and technology students was considered to be imperative. The demands of the industrial sector for increased quality and quantity in higher education in science and technology were further accelerated. All of these emphasized the necessity of specialist education in universities in order to meet industrial demand.

Following this increase of 8,000, another plan for expansion was begun in 1961, since the plan for doubling the national income announced in that year called for 170,000 more practising scientists and engineers over the period from 1960 to 1970, in addition to the projected number based on the then current levels.[20]

Accordingly, in 1961, the Ministry of Education launched a plan to increase enrolments in science and engineering departments of some 35,000 by an additional 16,000. This figure was reached under the assumption that demand would increase gradually every year during the 1960's. An increase of 16,000 students would cover the gap of 170,000 which existed between supply and demand in four years.

The demand for scientific manpower at the time was still higher than the Ministry of Education had estimated. The Ministry soon revised its estimates and announced an increased target number of 20,000 science and engineering students in September 1961, to be achieved by 1964. This was actually attained in three years and the entrance quota of science and engineering students reached 51,140 in 1963. More than half of the second increase plan of scientific manpower was covered by private universities.[21]

Abandonment of Egalitarian Policy: Higher Technical Schools

One of the major features of postwar educational reform was, as mentioned earlier, the promotion of egalitarianism as exemplified in the change from a plural to single track education. Such an idealistic approach was challenged by the industrial sector, which demanded changes in the egalitarian system to meet its immediate need for scientific manpower. The view expressed by the industrial sector ultimately led to the establishment of higher technical schools to train lower class industrial scientists and engineers.

In April 1961, a bill for the partial amendment of the School Education Law was presented to the 38th Diet. Enacted into law, it resulted in the inauguration of higher technical schools (junior college standard) in 1962.[22] In the preceding year of 1960, the Japanese anti-American, anti-establishment student movement against the renewal of the US-Japan Security Treaty had come to a peak. To avoid such ideological 'contamination' of their prospective employees, the industrial sector wished to have a school to train those students who would be obedient to company hierarchy, technically-minded, and untainted by university students. The graduates of these schools were to be treated more favourably in employment than college graduates. At least at the beginning of the system, there were more applicants than the Ministry of Education expected.

The industrial sector was always short-sighted in fulfilling its immediate needs for scientific manpower. When economic recession came, many graduates of higher technical schools were unable to find employment in their chosen field. The number of applicants for higher technical schools declined to half of the original number in 20 years.[23] The industrial sector also discounted the aspiration and upgrading trend at the grassroots level for higher education. More students preferred to go to universities, a desire which was met by the newly established science and technology schools of private universities.

The Role of Private Universities

Japanese universities were created within the public sector in the late nineteenth century, but as the government could not catch up with nor finance the increasing demand for higher education amongst the population, after World War I it gave university status to private schools. Thereafter, private universities developed to the extent that, just prior to World War II, the number of their students was twice the number at the public sector universities.

Higher education in science and technology, however, could hardly be maintained in the private sector, as the facilities are far more costly than for the humanities and social sciences and hence cannot be covered by the revenues from student tuition fees. Hence, since the birth of modern universities in Japan it became a tradition for the government to be the sole sponsor of science and technology training. During the prewar period, a few major private schools such as Doshisha, Waseda and Keio, sustained science and engineering schools but they were to be the exception. The overall prewar picture was that science and technology was a state-sponsored activity not only in education and training but also in research.

The Ministry of Education in responding to the late fifties demand from the industrial sector, since it could not increase the public sector budget so rapidly, turned to the private sector and encouraged private universities through the granting of subsidies and charters. This policy is somewhat parallel to MITI's science policy, if, indeed, MITI had a policy.

The first response was the Special Subsidy for Private University Science Education which was commenced in 1956. This was a subsidy amounting to 50 percent of the cost of equipment for student use in science schools. To this, a new subsidy was added in 1958 to further the plan for fostering science and technology education. It was designed to provide aid for further expansion of the schools of science and engineering. The aid which was provided was on a two-to-one matching basis, the school providing one-third of the total. Consequently, the number of science students in private universities came to outnumber those in public universities (Fig. 4).

The subsidy at first was limited to universities, but the science departments of junior colleges were included in 1959, followed by the newly established higher technical schools in 1962. This type of assistance has increased every year.[24]

In instituting new schools and increasing the entrance quota, public and private universities have been provisionally (at least for the first four years) required to consult beforehand with the Minister of Education, who in turn consults their University Chartering Committee (abbreviated to UCC), consisting of old professionals. The UCC initially tried to strictly maintain the old criteria of qualification and often intimidated the proponents of university reform. The UCC was particularly strict on private universities with dubious economic provisions.

In 1962, this procedure was largely mitigated to allow expansion of the entrance quota. The reason for the change was due in part to the fact that the Director of the Science and Technology Agency (STA), in connection with the plan for training scientific technicians to assist in carrying out the program of doubling the national income (March 1961), advised the Minister of Education to take this step. Normally, the sectionalistic practice of Japanese

Figure 4 Statistics of science and technology students in private and government universities

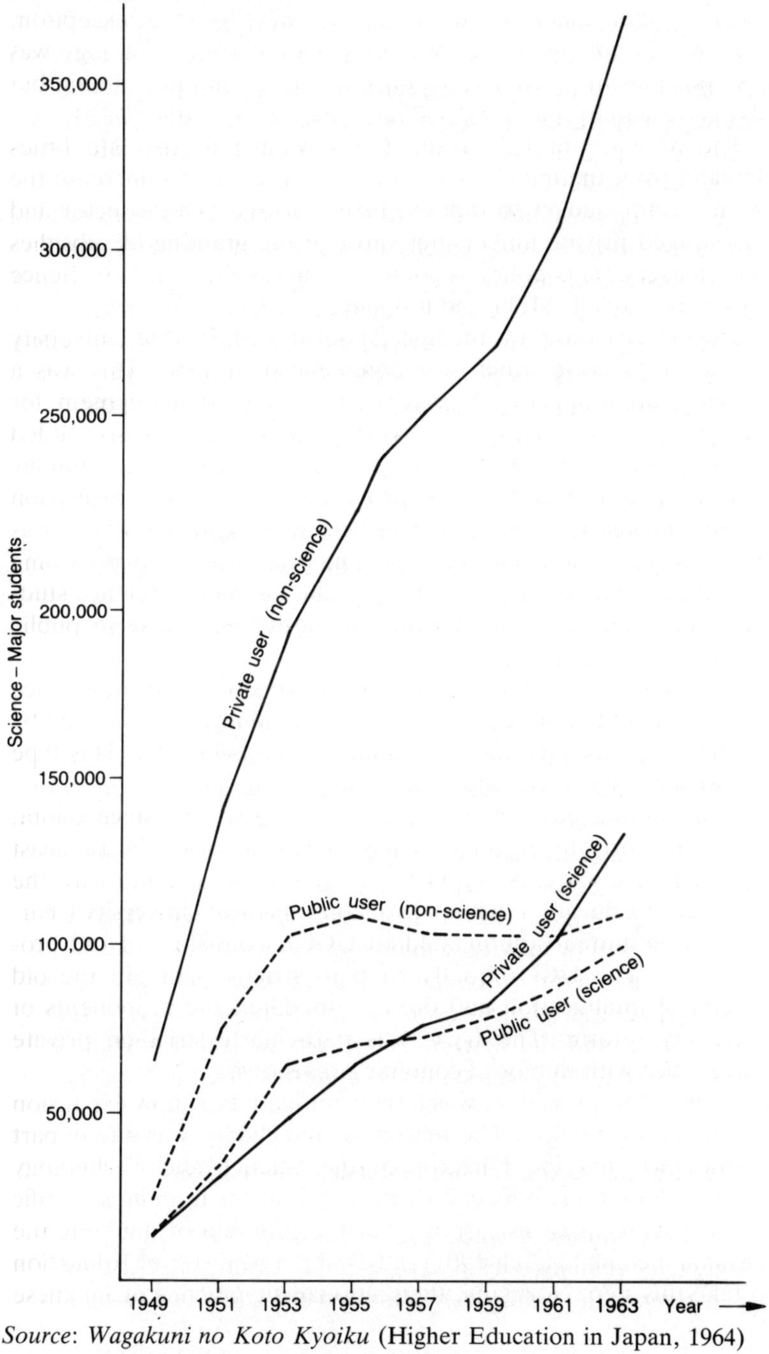

Source: *Wagakuni no Koto Kyoiku* (Higher Education in Japan, 1964)

bureaucracy makes it impossible to interfere or advise on the conduct of other bureaus. The top ministerial level is an exception. At this level they are not career bureaucrats but politicians of the ruling party (Liberal Democratic Party). In other words, only through this top level channel, can STA influence Ministry of Education policy.

From that time and until 1976 when stricter rules were returned to, universities, relying chiefly on their own judgment, could far more easily increase the number of schools and students, by only notifying the Ministry of Education rather than obtaining the latter's permission. In this period, the expansion of private universities was remarkable,[25] to a degree far greater than the Ministry of Education planned. The expansion of private universities contributed to the further cheapening of the quality of Japanese universities as the student-teacher ratio in private universities is three times higher than that of public universities, whilst tuition in the former is twice as high as that in the latter.

In addition to government subsidies, contributions from industries were made available from early in the 1960's to help to develop science schools. Some of the schools at private universities were in transition from professional schools (junior college standing) established during the war to meet wartime demand, while others had been attached to large corporations.

Most of the private universities are permitted to create postgraduate courses, as the government is under no obligation to share in the costs. If the increase of science and engineering schools and postgraduate courses in private universities has not contributed to academic science to the same extent as older national universities have, they have neverthless influenced economic growth by providing middle-class engineers to the industrial sector. Less favored are the biological sciences, and, especially, the agricultural sciences owing to the modernization of Japanese agriculture and the conversion of agricultural departments into industrial ones (e.g. agricultural chemistry, food processing, etc.).

The Realities of the Science and Technology Boom

The Post-Sputnik science and technology boom now makes for legendary reminiscence among elderly American scientists. Their first topic is the luxuriously allocated R&D expenditure, and secondly, the expansion of graduate schools. The Japanese boom did not appear to be due to a dramatic rise in R&D efforts, but mainly was a result of the rapid growth of scientific manpower to serve the industrial sectors.

For individual scientists and engineers, the boom does not necessarily coincide with the rapid rise of their status and work conditions. The increased power of the scientific sector may be compensated for by the loss of rarity value of each member. Science and engineering graduates are more able to find employment at times when they are in demand, but once an economic recession arrives, graduates (such as those of higher technical schools) experience hardship.

Such being the case, the science and technology boom was promoted mainly in public and private sectors for their own purposes. The citizen sector, however, does not necessarily respond so readily. High economic growth has certainly resulted in high demand for higher education on the part of the grassroots. While public schools were designed to meet the demand of the industrial sector to increase science student enrolments, private universities, which respond to the market mechanism, increased non-science students as well. In spite of public and private efforts to lure students into science and technology, many turned their back on science in preference for the humanities and social sciences (Fig. 4).

Notes

1 This is especially so in the cases of former Japanese colonial universities in Seoul and Taipei. Sungho Lee, 'The Emergence of the Modern University in Korea' in P.G. Altbach and V. Selvaratnam eds. *From Dependence to Autonomy: the Development of Asian Universities* (Kluwer, 1989).

2 General Headquarters, Supreme Commander for the Allied Powers, Civil Information and Education Section, Education Division, *Education in the New Japan* (Tokyo, May 1948, printed from typewritten copy) p.273 ff.

3 *Higher Education in Postwar Japan* pp.9–10.

4 *Daigaku Kijun Kyokai Junenshi* (Ten Years' History of the Japanese Association of University Accreditation, 1957) pp.49–57.

5 At Tohoku Imperial University which was founded in 1907.

6 Shigeru Nakayama, 'The Role Played by Universities in Scientific

and Technological Development in Japan', *Journal of World History* vol.IX (1965) pp.344–356.

7 *Higher Education in Postwar Japan; the Ministry of Education's 1964 White Paper* (edited and translated by John E. Blewett, S.J., Sophia University Press, Tokyo, 1965), p.17. The original version is *Wagakuni no koto kyoiku* (Japanese higher education, Monbusho, 1964).

8 *Higher education in Japan*, p.37.

9 Nihon Kagakukai (Japanese Society of Chemistry), *Nihon no kagaku o torimaku kenkyu kankyo* (Research environment of chemistry in Japan, 1988), p.40.

10 This is particularly strong in the American educational community, which proposes to upgrade high-school achievement standards by focusing attention on the high achievement of their industrial competitor.

11 Ikuo Amano, 'Continuity and Change in the Structure of Japanese Higher Education', in W. K. Cummings, I. Amano, and K. Kitamura (eds.) *Changes in Japanese University; A Comparative Perspective* (New York, Praeger, 1978) p.38.

12 OECD, *Towards mass higher education*, (1974).

13 'Reorganization of Science and Technology in Japan', Report to the US National Academy of Science, issued in August 28, 1947.

14 Katsuhiro Arai, 'Training System' in 'Rikogaku daigakuin no kakushin ni kansuru seisakuteki kenkyu' ('A Policy-oriented Study on the Innovation of the Graduate Training in Engineering Contents'), *Koto kyoiku kenkyu kiyo* (*Bulletin of the Institute for Higher Education*), No.10 (1989) p.71ff.

15 Mitsutomo Yuasa, *Nihon no Kagaku Gijutsu 100 Nenshi* (Centennial history of Japanese Science and Technology, Chuo Koron, 1984) vol.2, p.512.

16 *Ibid*. p.8.

17 IDE.

18 *Higher Education in Postwar Japan* p.64f.

19 Yearly statistics of student numbers are available in *Higher Education in Postwar Japan* pp.154–155.

20 The estimate varies according to the definition of scientific manpower. The 1960 national census estimates 330,000 while the Ministry of Education 800,000 and Nikkeiren (Association of Japanese Managers) over a million. *Kagaku gijutsu hakusho* (Science and technology whitepaper, 1965) p.239.

21 *Higher Education in Postwar Japan*, pp.65–67.

22 *Higher Education in Postwar Japan* p.66.

23 *Kyoiku Nenkan* [Education Yearbook, 1985].

24 *Higher Education* p.84–85.

25 *Higher Education in Postwar Japan* pp.25–26.

4 Expansion and Limit of Academic Science

The relative status of Japanese academic science vis-a-vis private science has undergone gradual decline over the postwar period. We trace, in this chapter, the effort made by the Ministry of Education to promote academic science and their subsequent problems.

Grants-in-Aid of the Ministry of Education

The university-chair system was introduced into the Imperial University in 1893 and extra funds for each chair was allocated mainly in order to encourage research in assigned disciplines.[1] Faculty members are assigned narrow, unchangeable duties which are prescribed to their chairs. Theoretically, the mandate attached to the chair binds the chairholder to the same research specialization for life.[2] Once assigned and formally established, the system functioned to discourage interdisciplinary research.

In the postwar period, the Occupation Forces attempted to abolish the system; however, it survived as a unit of budget allocation. The chair budget, once personal revenue for the chairholder, became the source of funding of all normal chairrelated activities, such as the education of students, the acquisition of equipment, the payment of utilities bills, and so on. New national universities created under the new system that the Occupation Forces introduced did not function under the old chair system and were allocated far smaller school budgets than the older, chaired universities.

As a normal school budget covers only the minimum maintenance requirements, a university scientist is compelled to apply for the grants-in-aid from the Ministry of Education. The scheme was created right after World War I on a small scale and, after World War II, it absorbed several other similar grantsinaid of the government to become the Monbusho Kagaku Kenkyuhi (Scientific Research Fund of the Ministry of Education, SRF), which is a major source for the promotion of academic science.

The system of fund allocation involves independent assessment by the scientific community so as to minimize any reasons for complaint. The major line of allocation policy is, however, determined by bureaucrats and the academic establishment. It encourages group research projects rather than those by the individual. The probability of acceptance of any one application is one in two for group research and one in five for individual research.

The Ministry of Education prefers supporting clearly defined research projects that are carefully scheduled rather than providing general research funds to an institution as a whole and allowing the money to be spent freely by its members. Thus the funding of the SRF increased dramatically whilst institutional research budgets remained constant. University scientists, who could not obtain the SRF, reflecting the egalitarian principles of the Chidanken, complained and for some time challenged the authority of the Ministry of Education. They could not successfully resist this trend toward the projectization of research.

The project funds are, however, allocated under the general account of the government budget. The government thus assesses the funds for projects in terms of scrutiny on their accountability rather than on scientific achievement. While research has the advantage of flexible funding, the SRF is subject to general administrative red tape, that hinders scientific activity. Furthermore, the accounting must be run on a yearly basis in line with the governmental budget. While the Japanese fiscal year starts in April, there are many bureaucratic procedures involved in the process of distribution and scientists gain access to the budget only in autumn and have to spend it by the end of winter. For example, if a scientist faithfully follows the government regulations of budget accountability, seasonal research such as that of agricultural science is quite impossible.

There are certain inherent limitations to the bureaucratic management of academic research, as the former requires accountability for the results obtained and rigorous financial assessment whilst the latter seeks the intellectual excitement of the unknown.

University-affiliated Research Institutes

In prewar times, Japanese university research was prestigious compared to that of private laboratories, because the industrial sector's research ability was not of a competent standard. In comparison with the structure of research in industrialized Western nations between the wars, where the troika of academic, public and private laboratories, had complementary functions, it was only university research in Japan which maintained higher standards and prestige than the activities of other sectors.

However, university research was, at that time, being conducted within the chair system, a small unit consisting of one professor, one associate professor and two assistants. It essentially developed and maintained merely as a unit of bureaucratic convention. It was criticized as an obstacle to the promotion of research. While the chair is often no more than a formality, it has survived to this day. Laboratory democracy movements during the postwar democratic period criticized the system and finally ignored it. Still, in order to carry out team research more effectively, a much wider research unit was needed.

It was for this purpose that a number of 'university-affiliated research institutes' came into existence after World War I. The first such institute was the Aeronautical Research Institute, affiliated to the Tokyo Imperial University, which was inaugurated in 1918. This was followed by the Tokyo Astronomical Observatory of Tokyo Imperial University (1921), the Metals Research Institute of Tohoku Imperial University (1922) and others. In these institutes, professors and assistant professors were all appointed to engage solely in research work, yet were extended the same academic status as university professors. The researchers also enjoyed academic freedom, academic prestige and freedom from teaching. They were certainly enviable positions unique in Japan, even from the viewpoint of Western scientists, who had generally enjoyed better facilities and working conditions than Japanese scientists.

University-affiliated research institutes have survived through the postwar university reforms, though during the Occupation period those institutes said to be involved in the war effort were largely reorganized. Academic autonomy has allowed them to escape scrutiny in the postwar reorganization of the public sector. Though the academic sector experienced considerable economic hardship after World War II, academic science for its own sake was relatively prominent during that time, compared with the mission-oriented technocratic science of public and private laboratories, which were unable to organize themselves adequately and identify their research goals amidst the postwar devastation and confusion.

Common Use Research Institutes (1953–)

The recovery phase of the postwar economy in the 1950s saw growth in the capitalintensiveness (or resource-intensiveness) of academic science. As the scale of research projects expanded, a need arose for collaborative research by university scientists using large and expensive installations. Postwar laboratory democratization led to the call by the scientific community for equal participation in largescale projects and more shared use of resources. It was no longer possible for one particular university to monopolize all the limited resources of a particular field.

In order to avoid laboratories duplicating facilities, a centralized 'Research Institute for Common Use' system was created with help from the Japan Science Council, instead of the then existing system of university-affiliated laboratories. After the establishment of some of the earliest examples of this type of joint use institute such as the Cosmic Ray Observatory and the Research Institute for Fundamental Physics founded in 1953, large well-equipped research institutes were established for the common use of scientists to enable them to carry out top-level research.[3]

Most Japanese universities that have a research function in science have been government universities, in which all members, professors and assistants alike, are civil servants, with tenure and the protection of employment up to the retirement age. Seniority rule and salary scale according to age are also characteristically bureaucratic. Sabbatical leave, unique to the private academic system, has never been granted to the civil servant faculty members of public universities.

There is another aspect of the bureaucratic model of Japanese universities: the role allocation in each compartmentalized department is thoroughly and rigidly maintained. The organization of students also follows the organization of faculty in departments. Once students are enrolled in a department and supervised by a professor, they are not permitted to change their supervisor. Similar to the inter-ministry walls that exist in civilservice administration, once committed to this kind of organization, there is, in principle, no horizontal mobility, unless one revolts against the department common law. At the inter-university level, there is practically no exchange of personnel. Inbreeding was rife.

Postwar laboratory democracy attempted to eradicate the inter-laboratory and inter-university walls and were successful to some extent. Its achievement, the establishment of common use laboratories, became the centre of interuniversity exchange. Thus, democracy is more prevalent in laboratories than in educational departments; in a certain sense, a single university hierarchy does not operate in laboratories.

Big Academic Science – Case of KEK

Democracy and centralized big science are hardly compatiable. Since the democratic consensus of members of a scientific community is essential in promoting a big project, the conclusions tend to be conservative.

During the Occupation years, nuclear physics research was prohibited by GHO, while it dramatically advanced elsewhere. Taking advantage of the 1953 Eisenhower call for 'atoms for peace' and the Nakasone budget increases in the following year, some leaders of the Japanese nuclear physics community tried to move towards the world standard urging the establishment of large accelerators in Japan. This was the beginning of Japanese big academic science but the infrastructure to promote such a big plan was economically as well as institutionally immature.

When a proposal was made by nuclear physicists to build a 1 GeV sychrotron at the Institute for Nuclear Studies affiliated to the University of Tokyo in 1955, it was bitterly criticized by a majority of the physics community, as they feared that the funds for their own research would be cut off. This was a serious problem for them, as the research funds of the Ministry of Education are allocated, in principle, so that each field gets a sum in proportion to the size of its membership.

Long debates were held on different proposals from scientists or groups of scientists within the Japan Science Council. These debates involved the wider scientific community so that they considered deeper political problems arising from questions concerning the government's commitment and the scientists' autonomy.[4]

While the debates were going on, the accelerators at Brookhaven and CERN were scaled up so that the originally planned 1 GeV accelerator became obsolete by any standard. Thus, the Japan Science Council, in 1962, adopted the proposal for a 12 GeV proton synchrotron in a new laboratory site. Almost another decade was lost, however, with further discussions on the scale of the budget, effects on and relation to research programs in other fields of science, and the organizational problems involved in the establishment of a new national laboratory.

Finally, the High Energy Physics Research Institute (Ko-Enerugi Kenkyusho, the KEK) was established in 1971 at Tsukuba with a reduced budget and plan, and the first instrument of the KEK, a 12 GeV proton sychrotron, was completed in 1976.[5] At the inception ceremony, those concerned congratulated themselves reflecting on two decades of debates, and commented that 'we are not yet sufficiently mature to challenge the international frontier of research but we have learned a valuable lesson in how to establish a big science laboratory.'[6]

NIURI – a National Institute (1971–)

In a democratic organization such as the Japan Science Council, it takes a long time to reach a decision, especially when there are political problems associated. After these events, the government turned their back on the Japan Science Council, whose recommendations for the creation of laboratories so far accepted, in preference for the newly appointed Science Council (Gakujutsu Shingikai, established in 1967) of the Ministry of Education (Monbusho, which later changed its English title to the Ministry of Education, Science and Culture, MESC). Although the government's unfavorable attitude towards the Japan Science Council is believed to be due to ideological differences as noted in chapter 2, it is also the case that the scale of academic science has now developed to a level beyond which the academic sector could manage by itself.

When the establishment of KEK was authorized in 1970, it was directly affiliated with the Ministry of Education as the first NIURI (National Inter-University Research Institute), though professorial status of members remained unchanged. Until that time, Common-use Research Institutes were, as a matter of administrative convention, managed by a single university which was responsible for their coordination. The NIURI laboratories not only exceed the scale of singleuniversity management, but are also open to users (academic as well as industrial) and international collaborators. As a result, an academic science laboratory can find potential clients in the industrial sector. An accelerator itself can be used for the development of new technologies and new products for industrial use.[7] Big academic science in Japan has gone on to attract cooperation from overseas scientists.

After 1968, university research became adversely affected by student revolts. University faculties and administrations were preoccupied with student problems and paid less attention to research. University-affiliated research institutes were less affected, with most of them being located outside of the campus. They found little merit in affiliation with the universities.

In this social context, other big academic sciences followed suit. The National Institute of Polar Research (NIPR) in 1973 and The Institute of Space and Astronautical Science (ISAS) in 1981 became members of NIURI.[8] ISAS and the Tokyo Astronomical Observatory, both formerly affiliated to the University of Tokyo, became directly affiliated to the Ministry of Education in 1981 and 1988 respectively, as their budget became a burden too large for a single university to manage. As long as they were affiliated to the university, their proposed budget would have been curtailed by other jealous institutes and departments within the university. In order to launch satellites for space research and to obtain the huge budget needed to start a new high-tech telescope, these institutes could not be restricted by the normal allocations of the scientific community, where the size of budgets is democratically determined.

Thus, though university academic science is impoverished, NIURI institutes have gained access to a big science budget appropriate to Japan's economic strength. On the other hand, some critics entertain misgivings about the curtailment of academic freedom, once the institutes are subject to direct bureaucratic control. Furthermore, the weakness of national institutes, distant from universities, is that they face problems not only in recruiting manpower but also in gathering fresh ideas. During the riot period of the late 1960s and early 1970s, old faculty members suffered from demands for laboratory democracy made by graduate students and were happy to get rid of graduate students at the NIURI style institutes. But it is often found that a new research institute can be quite active in its first decade but later lose intellectual vigour unless it successfully imports new blood in the way of young graduate students. For this reason it has been proposed that the relation of NIURI to graduate schools and graduate students be improved.[9]

It should also be noted that big academic science under the auspices of MESC is still governed by the evaluation criteria of academic science. Thus, both KEK and the Institute of Space and Astronautical Science pursue their own academic goals, without being much involved in national public service.

National Prestige and International Cooperation

Looking back at the scaling up of academic science laboratories from the chair system, to University-affiliated Research Institutes, to Common-Use Research Institutes and finally to NIURI, we are tempted to extend the scaling-up beyond national barriers to international cooperation. Japanese leaders of big academic science envy and often cite the successful attempts of the European Community's CERN.

Academic science has been motivated by individual competition, while public and private sciences by international and interfirm competition respectively. For the promotion of international cooperation in academic science, something to replace competition must be found. A number of problems in the infrastructure mechanism need to be solved, such as how scientists persuade participating governments in allocating funds for cooperation.

There is a fundamental question in the legislature and administration which deals with the rationale of funding academic science at the tax-payers' expense. On the funding of big academic science, argument becomes volatile. Leaders of big science must find an adequate explanation in order to persuade the Ministry of

Education, then, in turn, the Ministry of Finance and finally the Diet. When a big science proposal for the installation of a 76-inch reflector was made for the first time in the postwar period in the 1950s, two methods of explanation were employed: national prestige and international cooperation.

Astrophysical and cosmological use of the telescope is highly centralized. The largest telescope, a 200-inch reflector on Mt. Palomar, makes any smaller one academically meaningless. Against this academic argument, proponents of the plan claimed that 'Japan was terribly behind in installing a big telescope. The largest one we have is only a 26-inch Germany-made refractor. If an international visitor were to ask for the largest telescope in Japan, the Japanese would be ashamed. Hence, no matter what its academic use, we have to have a proper instrument of a size appropriate to our national power.' This type of argument had a considerable persuasive power on the Diet members.

Another explanation is the international cooperation (division of labour) argument, in which Japan's geographical location was emphasized. Because of the longitudinal difference, at the time when astronomers of the USA and Europe were unable to observe the night sky, the Japanese can. Unlike positional astronomy, the exclusive use of a big reflector for astrophysics does not necessarily validate this argument but it is still a more academically convincing one. The latter argument can be also traced back to national prestige, as it would provide Japan with the opportunity to participate in the international research frontier. This type of argument applies to other big academic science also.

When Japan was still in the development stage, international cooperation for Japanese scientists meant learning and catching up with advanced science even at great national cost. When that stage was over in the 1970's, and to an even greater degree in the 1980's, benefits from participation in international cooperative programs became less significant. At the same time Japan is now expected to pay its due share of the financial expense of international projects as the 'duty' of an advanced nation.

The question of how one can transform feelings of 'duty' into the accountability of a national budgetary system is of great concern to the public sector as well as the academic sector, though the reason of concern is very different between the two sectors.

International cooperation with advanced countries often excludes the participation of scientists of less developed countries.

In this case, alternative types of feelings of duty must be addressed to the less developed surrounding countries. Academic scientists in universal science disciplines until the 1960s, who had been solely preoccupied with catching up with the West, had practically no concern for this aspect. Very few physicists ever visited even the neighbouring country of Korea.[10] Only a handful of concerned physicists have attempted to organize inter-Asian meetings and scientists have so far failed to find a methodology of 'appropriate science' comparable to 'appropriate technology'.[11]

The Difficulty of Cooperation between Academic and Industrial Sciences

The MESC has been the sole public sponsor of academic science. Its aim has been, in simple terms, the creation of more Japanese Nobel laureates. This aim has not been very successful so far, compared to MITI's international prestige in supporting and promoting private science. This is in spite of the fact that MESC spends nearly half while MITI only 13% of the government budget for science and technology (in 1988).[12] Perhaps the only consolation for frustrated MESC bureaucrats is that big academic science, the support of which they are responsible for, cannot be evaluated in terms of the number of Japanese Nobel laureates, as capital-intensive group research is not the major recipient of the Nobel prize, which was begun principally to encourage individual small science at the turn of the century.

The bureaucratic model of Japanese universities discourages academic cooperation with industrial laboratories. The government employee's rule has been applied to national university professors, who are almost forbidden to cooperate with private enterprise. The second hindrance is the anti-war sentiment prevailing in the academic sector, where even pragmatic engineering professors are influenced by idealistic students and social scientists on their campuses.

The student body, led by leftist leaders, has always resented external pressure and identified it with a capitalistic, and symbolically militaristic invasion of academia. This was persuasive in light of the 'Red Purge' in academia carried out by the Occupation Forces during the Korean War. Consequently, Japanese universities could successfully resist outside pressures from industry and remain aloof from cooperation with industrial and particularly military research, as Harry C. Kelly, an advisor in science and technology during the Occupation, concluded in a comparison with American universities.[13]

An undercurrent of pressure against the professors who had cooperated closely with industry gradually built-up and finally erupted during the campus unrest during the transition period from the rapid industrial growth period of the 1960s to the depressed period of the 1970s.[14] Since then the subject of university-industry cooperation has been a taboo throughout the 1970s for those engineering professors who were intimidated during the university protest period.

The third impediment arises from the industrial view of academic science. During the period of importation of technology after the War, most of the technologies used by Japanese industry were based on those invented and developed in the United States and Europe, whereas the academic science of engineering aimed for international recognition abroad. In other words, industrial scientists see academic science as too remote, whilst academic scientists in engineering find the expectations of industry too short-sighted. Distrust and the motivational gulf between the two camps are not easily eradicated.[15]

In the meantime, haughty Japanese universities have turned out to be obsolete in equipment and facilities for scientific research because of their isolationist attitude. Even in academic science, research facilities are becoming increasingly expensive and no longer affordable by university budgets. Highly expensive computers and other facilities are often donated by a company, which then takes advantage of the respectable university as a show-case for their equipment. It is reported that the research budget per capita of researchers at a national university laboratory amount to somewhere between 1/15 to 1/40 of that of private laboratories.[16]

Some concerned university scientists have, in the late 1980s, attempted to promote university-industry cooperation, in view of the poverty of academic science. In spite of student opposition and the apathy of non-scientific faculties, the University of Tokyo tried to introduce industry-endowed chairs in 1988. Industrial scientists are not terribly enthusiastic about this practice, although they do not explicitly oppose it. A leading industrial scientist boasted that 'we do not have to depend on university research at all. If we need basic science, we can do it by ourselves.'[17]

Academic Scientists Abroad

Before the War, university faculty members had better opportunities to study abroad than members of industry, who had practically no opportunity to do so. As a result, the universities became national centres of research and development and the gateway to foreign technologies.[18]

After long wartime seclusion, the first Japanese scientist who was invited to the United States in September 1947 was Hideki Yukawa. J. Robert Oppenheimer, the director of the Institute for Advanced Study in Princeton, had used his strong Washington connections to get a travel permit for him from the Japanese occupation authorities.[19] He was followed by Sin'itiro Tomonaga and other physicists.

In 1950, an official channel of sending students to the USA, the Garioa-Fulbright program, was initiated. Besides this, there were a number of academic scientists and engineers in the 1950s who found their way to the USA through professional channels. Many of them were hired as laboratory staff or research assistants financed by various types of research funds. Though an established Japanese professor was often employed as a cheap labourer, many of them found a new paradigm emerging abroad that was not known in postwar Japan, returned home and became leaders in the new disciplines. In contrast, few Japanese industrial scientists were to be found on American campuses.

In the post-Sputnik science boom in the USA, the brain drain was widely discussed in the scientific community, as a great many scientists were being lured to the USA with attractive offers. Indeed, the Tokyo Astronomical Observatory was at one time nearly evacuated as its staff members were, one by one, scouted out by NASA.

However, the brain drain did not occur in Japan to the same degree, as it did elsewhere[20]. Nearly all of the Japanese who were invited overseas returned home after a few years. The major reason for the absence of brain drain is that Japanese academic scientists have, in principle at least, a tenured position at their home universities, which may not be casually and inconsiderately cancelled. In addition, the language barrier for the Japanese scientists is greater, in one measurement four times higher[21], than for European scientists. Though ambitious experimental scientists sometimes remained in Japan as the working conditions at home were superior to those overseas even though their standard of living was lower, other considerations, such as racial prejudice prevalent in some American communities and difficulty in educating and communicating with their child in the Japanese language, all serves to discourage expatriation of the Japanese.

In the 1970s, the situation reversed after the high economic growth of the 1960s. US academic science no longer had the funds to hire Japanese academic scientists since the severe budget cuts which have prevailed since 1968. On the other hand, the Japanese industrial sector began sending their staff to American institutions, to the laboratories where professors have research contracts with Japanese corporations. Thus, on the matter of information gathering, the relative impoverishment of Japanese academic science is evident in comparison with that of industrial science.

Bureaucratic Model to Business Model?

The bureaucratic nature – inefficiency, inflexibility, formality, lack of market accountability, etc. – of Japanese higher education has been criticized for more than a century. It has been under the control of the MESC since the founding of modern universities in the nineteenth century. MESC itself has tried to change into a more flexible and more adaptable organization to meet the needs of the business sector but its reforms are slow.

MITI and STA, with the push and support of the industrial sector, have on various occasions challenged the established pattern and expressed the desire to establish a science and technology graduate school of their own. They have, however, so far been unsuccessful because the MESC monopolizes the right to confer academic degrees.

In 1984, the Interim Council of Education (Rinji Kyoiku Shingikai) was formed as an advisory body to the Prime Minister. Up to 1987 the Council submitted draft reports in which various reform measures for the Japanese educational system were presented. Recurrent education, internationalization, flexible adaptation to the future information society, etc. are some of the major catch words in these reports. The reform of graduate schools has been largely modelled after the American system. Private contributions to national universities and NIURI in the form of endowed chairs has been allowed since 1987, in order to introduce private funds smoothly into national institutions to enrich and activate them.[22] Reform of the MESC is more easily done in the area of research rather than in science education. Since the mid-1980s, MESC has, after jealously viewing the success of MITI's strategy, promoted 'industry-academic cooperation'.[23]

Notes

1 *Teikoku Daigaku no Tanjo* p.140ff.
2 Yoshio Abe and Akira Tezuka, 'Perceptions on Basic and Applied Sciences in Japan', *Science Policy Perspectives* p.30.
3 *Higher Education of Postwar Japan* p.69.
4 I remember that during the debates Hideki Yukawa, the first Nobel laureate of Japan, privately talked to us (to the audience of the History of Science Society in Japan) in a critical fashion, remarking that 'something will certainly come out of a high-energy collision with a big expensive machine. It is a principle even a child can comprehend. It is not such elaborate science as we have been concerned with.' He was sorry that the 'good old days' of small academic science had died out, since big science is invariably entangled in technocracy.
5 Yoshiya Abe, Hiroshi Inose, Tetsuji Nishikawa, Kotaro Shimo, and Akira Tezuka, 'Policies and Practices of Government Support for Basic and Applied Science', in Arthur Gerstenfeld ed. *Science Policy Perspectives: USA-JAPAN* (Academic Press, 1982) pp.151–155 and L. Hoddeson, 'Establishing Fermilab in the United States and KEK in Japan: Nationalism and Internationalism in High-energy Physics', *Social Studies of Science*.
6 Talk of Professor S. Suwa, the first director of the KEK.
7 Hiroshi Inose, Tetsuji Nishikawa, and Michiyuki Uenohara, 'Cooperation between Universities and Industries in Basic and Applied Science' *Science Policy Perspectives* p.59.
8 List of NIURI members can be found in *Science Policy Perspectives* p.152.
9 *Science Policy Perspectives* p.169.
10 When I visited Korea in early 1970s, I was told by Korean physics and astronomy societies that the Japanese scientists so far had never been interested in visiting Korea.
11 Kagaku, ningen, shakai.
12 *Kagaku gijutsu hakusho* (Science and technology white paper, 1989) p.228. Although the absolute value of percentage is not a reliable indicator, as the definition of science and technology budget varies by each governmental section, it is interesting to note the recent

trend that MITI increased from 7 to 13 for last ten years, while the MESC and STA decreased from 54 to 47 and from 29 to 26 respectively.

13 Harry C. Kelly, Interview by Charles Weiner (1975). Institute Archives and Special Collection, MIT Archives.

14 Hiroshi Inose, Tetsuji Nishikawa and Michiyuki Uenohara, 'Cooperation between Universities and Industries' *Science Policy Perspectives* pp.45–48.

15 *Ibid.* p.49.

16 'Rikokei Daigakuin' p.12.

17 Personal communication with the former director of Hitachi Central Research Laboratory.

18 Inose *et al* 'Cooperation' p.47.

19 *Science* 23 May 1949.

20 The census of foreign scientists working in the United States conducted by the National Science Foundation does not put Japan as an item in a list of some forty countries.

21 In one estimate, Japanese scientists spend four times more time in writing a scientific article in English than European non-native-English speaking colleagues.

22 *Wagakuni no bunkyo shisaku* (Japanese educational policy, 1968, MESC) p.169.

23 Ibid. pp.173–174.

5 High Economic Growth and Private Science

Postwar Japanese science is characterized by the supremacy of private science. In this chapter, we try to ellucidate the uniqueness of its character and explain how it came into being.

Miracle Explained

Japan's postwar economic growth is often labelled by outside observers as a 'miracle'. Such an irrational term is a challenge and, indeed, unacceptable to rational historians and social scientists.

Some attempts by outside observers have been made to explain the cause of Japan's high economic growth in terms of economic policy.[1] Their analyses are often motivated by the hidden intention of adopting Japan as a policy model.[2] The present writer, as an insider, does not share much of their modelseeking problematique.[3] Instead, an insider may be able to give a critical assessment focusing more on the nature and quality of Japanese scientific and technological manpower.

On the other hand, there is not much explanation made in terms of science and technology policy, perhaps because if one looks at it, one will be impressed by the significant absence of a thoroughly planned science policy in postwar Japan, as compared to those which have existed not only in socialist countries but also major industrialized countries in the Western world. Nevertheless, the cause of Japan's high postwar economic growth is, as most writers agree, largely due to technological advancement. In the following, we shall try to solve this riddle mainly in terms of scientists and technologists in the private sector who are involved in the course of events.

Historical Origin of Modern Japanese Technocratic Scientists

In the mind of an historian of Japanese technology, two major, overlapping historical recollections of Japanese experience can be found: rapid Westernization of the late 19th century and the postwar 'miracle' as seen by outside observers. Those who have studied the former experience well enough cannot be easily persuaded to call the latter a miracle, as the former one appears to be equally or perhaps even more miraculous. It will be unfair to history to explain the latter 'miracle' without mentioning the earlier one, as it is a continuous endeavour of scientific and technological manpower of the same type, not replacing one group with another.

The Samurai Origin of Japanese Technology

In the West, as well as traditionally in the East, the practitioners of technology came from a relatively low social class. Modern Japanese science and technology professionals, on the other hand, came from the upper class, the samurai, comprising the top five percent of the total population.

These modern professions were the product of the new Western-oriented Meiji government. Thc main practitioners were former samurai, who were warriors by definition but who, during the Tokugawa period, became primarily administrative bureaucrats. In the past they had received hereditary family stipends.

In the 1870s, efforts were made by the Meiji government to curtail the inherited stipends as a preparation for modernization. Consequently the samurai completely lost their source of revenue and were forced to find an entirely new means of living.

One of the government's modernization programs involved the rapid absorption of Western technology. As early as the 1870s, the government established and conferred university status not only on science but also engineering, without being overly influenced by the Western university tradition. Science and technology were among those new fields into which jobless samurai were attracted and invited. It was reported that almost all of the early graduates of the school of engineering were samurai.[4]

Samurai were the class long accustomed by mental habit to thinking in terms of public affairs and, by behavioural pattern, to playing the game of public office. Science and technology were, at least in the early Meiji period, purely governmental enterprises which were given top priority. The samurai were naturally attracted by government-sponsored careers in science and technology. They found their most promising career possibilities in the entirely new professions of technocrat in the building of a modern state[5] Very few of the samurai were oriented to academic science, an area to which social status had yet to be attached.

The Dual Structure of Japanese Technology

Modern engineering and technology may be said to have two entirely different origins. One is the 'public-centered' engineering service practised in the public sector, best exemplified by the discipline of military and civil engineering which is taught and practised in the celebrated French ecole polytechnique, and the other is 'private-centered', profit-making engineering practised in the private sector, as exemplified by the Watt-Boulton type of enterprises, mechanical and electric engineering, pharmaceutical technology and so on. The former may be defined as 'public technology', while the latter can be called 'private technology.'

The above dichotomy has arisen in the Japanese technological world in a more pronounced manner than even the Western precedent. Public science and technology (or government-sector science, or non-industrial technology) is, as noted above, not only 'public-centred' but of Western origin, university-based and practised mainly by the 'policy-oriented' samurai class elite. On the other hand, private technology is not only 'private-centred' and profit-making, but also traditional and domestic, and transmitted through apprenticeship in the non-samurai sector, mainly by the traditional class of craftsmen, in areas such as carpentry, sake brewing and ceramics.

In the early effort in the 1870s, the government attempted to introduce both public and private technologies together indiscriminately. Elite graduate engineers who were busy introducing and translating the high technology imported from the West, planned and designed a nationwide project of establishing a modern technological infrastructure such as a telegraph system and land surveying. They felt at home, in public works of a non-profit-making kind. They were, however, perhaps poor performers in profit-making kinds of private technology. Their technology often proved to be failures in terms of commercial accountability when applied to the experimental government factories and mines. This step was a valuable lesson for the industrialization carried out by the next generation.

Private technology was better carried out by craftsmen in the private sector, as best exemplified by Sakichi Toyota, the founder of the Toyota Corporation, and the son of a traditional carpenter. Though the craftsmen remained outside governmental enterprise, their inventive skill in the private sector was demonstrated in local technological adaptation or in appropriate technology[6]

While college graduates were employed by governmental reviews boards for domestic prizes and patents in the last quarter of the nineteenth century, they were not interested in obtaining patents for themselves[7]. They looked down upon inventors in the private sector and viewed them as illiterate, uneducated and unreliable. The agenda for themselves was the systematic adoption and improvement of technology of Western origin.

From the 1880s, the government transferred government-operated factories to the private sector in order to achieve economic accountability. Along with the privatization policy, proud samurai technocrats reluctantly moved to private enterprises. Vanguards of the public-to-private transfer were graduates of electric engineering. This new private technology was purely Western and hence no traditional craftsmanship existed to respond. Early graduates were sent abroad to undertake apprenticeships as technicians rather than to study at universities. Upon returning home, they produced Japanese electrical appliances, some of which became the base for the later microelectronic industry.

Since 1890, the government favoured college graduates from law rather than engineering. Two decades of effort prior to that in promoting Western science and technology was to some extent satisfied; the government then turned from the hardware to the software of maintaining a modern state, constitutional monarchy and modern efficient bureaucracy. Law school graduates of Tokyo University were assigned this task. Engineering graduates complained of the decline in their status relative to law graduates.

They campaigned for a levelling of promotion within the bureaucracy in the 1910s and as a result were successful in gaining positions in public technology-oriented offices such as the Ministry of Construction, where civil engineering graduates attained parity of status with law graduates.

The technology transfer from public to private sector was a protracted process. Proud samurai did not mingle easily with private-sector inventors even after World War I.

However, if we look at statistics of the number of employees in public sector factories, we can see that it was a process of constant decline for a century, with the exception of the war-time mobilization, ending only in the postwar period, when private sector technology dominated science and technology fields. The above brief history accounts for the relatively high social status of Japanese engineers.

Evaluation of Wartime Effort

In the historiography of science and technology of postwar Japan, it is an established thesis that Japanese science and technology developed with an imbalanced military orientation since its institutional modernization and Westernization at the time of the Meiji Restoration in 1868.[8]

Science mobilization during the war, is assessed as scientifically 'bankrupt' because of the lack of coordination and the dominance of ultranationalistic ideology.[9]

This historiography is undoubtedly biased or at least coloured by the pacifist anti-militaristic sentiment prevailing in postwar democratic circles, such as *Minka*. Later generations started to question to what extent this thesis holds on the basis of unbiased historical scholarship.[10] Recent studies also shows that the area where technology transfer from the West was successfully conducted in the early postwar period was identical with the area where prewar and wartime Japanese development was linked with imported technology. As to the early history, Chikayoshi Kamatani recently emphasized the effort made by the government to promote non-military, industry-bound national research projects in the late 19th and early 20th century.[11] Whether militaristic or non-militaristic, Japan was a country that started a technocratic structure earlier than any of the Western nations in the nineteenth century when science and technology was not yet necessarily linked to national endeavour.[12] In other words, it was plan-rational and developmental.[13]

More relevant to postwar development is the evaluation of the wartime effort. An examination of the growth of the Japanese scientific community reveals a lacuna created by isolation from Western scientific information during the war. It was some ten years after the war before Japanese science could recover, during which time the Japanese worked on imitating the technology of other countries and catching up to American science and technology. A few cases such as shipbuilding and optical industries were exceptional as the basic technologies were supplied by the transfer of technologies developed for military purposes during and before the war[14]

In terms of scientific and engineering manpower, prior to and during World War II, there appears to have been a kind of science and technology boom. Students majoring in science were exempt from or permitted to delay compulsory military service. Scientists and engineers were able to enjoy access to handsome research funds which they had never experienced before. Whether wartime scientific manpower contributed toward the postwar technological miracle in the 60s is an important question which should be addressed.

During the Occupation period of the late 40s and early 50s, military-bound science and technology was strictly prohibited in such fields as nuclear and and aeronautical engineering. Most of this type of industry was thus closed off to the Japanese. Scientists and engineers who worked hard to meet wartimc nccds, found themselves redundant after the war. Most of them abandoned science and engineering fields altogether out of frustration.

Military-bound university departments were reformed or disguised as peaceful disciplines; aeronautical engineering became applied mathematics or physical engineering, military technology was changed to precision engineering and so on. Some survived by slightly modifying their subject of study, from, for instance, aeronautical to automobile engineering. There were certain setbacks but structural reformation, namely 'scrap and build' operations, did take place.

Most notable was the dissolution of the Department of Aeronautical Engineering of the School of Engineering at Tokyo Imperial University, which, during the War, enjoyed the highest prestige and attracted the most talented engineering students. During the Occupation period, research in aeronautical engineering as well as the aeronautical industry were strictly forbidden. These young graduates of the Aeronautical Engineering Department had dispersed into various fields, such as railway engineering and automobile industry, where they brought new concepts of aeronautics into new areas.[15] The celebrated bullet train (Shinkansen) is one of the results.

The Occupation Policy and Economic Recovery

In the very early days of the Occupation, military personnel were assigned the task of making sure that no militarily strategic research was still being carried out by Japanese scientists. One of the results of this was the destruction of several cyclotrons – an event that created a furore among U.S. scientists, and was at least partially responsible for the staffing of the Scientific and Technical Division of the Economic and Scientific Section (ESS/ST) which, up to then, had no personnel.

Several of the GHQ civilian sections, even quite early in the Occupation, sought to find ways of promoting research that was judged likely to help with the economic recovery of Japan. This was particularly true of ESS/ST, despite the fact that they too were responsible for seeing to it that weapons research was not being carried out in any Japanese laboratory. Two advisory groups sent from the U.S. National Academy of Sciences to study the state of science in Japan reached the conclusion that, because of inadequate coordination of research, the findings of Japanese scientists were not being put to practical use in industry.[16] These American scientists were convinced that Japan had great capabilities in both research and industrial development, but that closing of the gap between the two should be accelerated. These viewpoints obviously received great emphasis in ESS/ST. H.C. Kelly, Deputy Director of the Scientific and Technical Division, was the most effective person in this respect. It was also received close attention and support from the economists of ESS.[17]

Rationalization of technology

In the meantime, in 1947, the Truman Doctrine of 'containment' of the communist world was adopted; accordingly, American policy toward Japan underwent a major change. Until this time, the primary role of the Occupation Forces was to serve as policemen: to see to it that Japan did not rearm and that reparation payments were made to past enemies. After the Truman Doctrine came into effect, Japan was turned into a stronghold against the communist offensive in Korea; to that end, the enhancement of productivity of Japanese industry was encouraged rather than discouraged, while the reparation burden was reduced. In this context of international politics, the rationalization of Japanese business enterprises was of course considered important, and this concept was pushed strongly by Joseph Dodge, the American banker and economist who was brought to Japan in 1949 to take charge of the program (mandated by Washington) to assure economic stabilization in Japan. Dodge pressed many measures on the Japanese government and financial establishment designed to accomplish his goal, including the abolition of government subsidies to key industries, and the rationalization of the managements of those industries.[18]

According to a renowned historian of technology, Yoshiro Hoshino, the critical turning point in the history of postwar Japanese technology was the arrival in Japan in 1949 of Joseph Dodge, and the ensuing rationalization of Japanese governmental management; this position is in contrast to that of most economic historians who believe the turning point to have been the outbreak of the Korean War in 1950.[19] The Dodge-inspired measures of rationalization were drastic, and many Japanese executives at first felt that these extreme measures were intended to destroy Japanese industries. This was not the case: the Americans simply wished to change the character of the Japanese economy, management and technology from premodern to a more modern style, based on the American model, in part so that they could procure Japanese weapons which met high standards for use in their confrontation with the communist forces.

Many chronological events prove that it was not after but before the breakout of Korean War when the Occupation Forces decided to change their policy from demilitarization to the development of Japan as a base camp of the Cold War.

The stern rationalization measures resulted in the reduction of research and engineering manpower as inefficient and redundant. As a consequence, many industrial laboratories (NEC, for example) were closed down.

Later in May, 1949 MITI was created, by reforming the old Ministry of Commerce and Industry and other ministries. As an advisory body to MITI, the Industrial Rationalization Council was organized. The Industry and Engineering Agency of MITI also published that year the first *Technology White Paper*.

Quality Control

Of the various lessons learned from the Americans during the Occupation period, the most appreciated by the Japanese was 'quality control'. The early study of quality control in Japan was limited to the academic world, where it was studied by applied mathematicians and statisticians.

Kelly and his colleagues at ESS/ST had encouraged the idea of using research as a means of assuring economic recovery since the early period of the Occupation – perhaps out of New Dealers' goodwill. But the general Occupation push toward assuring economic recovery only came in 1949 in connection with the severe international political realities of that time. In this context, Dr. Bowen Dees[20], Col. Allan, Dr. Dixon Reck, Mr. W. K. Kight and other members of the ESS/ST staff encouraged various groups in Japan to take up the concepts and practice of quality control,[21] but neither the Japanese public nor private sectors were very enthusiastic about putting their advice into practice.

In April 1950, *Nikka Giren* (the abbreviation of Nihon Kagaku Gijutsusha Renmei – the Association of Japanese Scientists and Engineers) commenced monthly publication of the *Hinshitsu Kanri* (Quality Control) journal. In July, they invited an American professor of statistics and industrial engineering, Dr. W. Edwards Deming, to come to Japan to hold a series of seminars on quality control for Japanese engineers. While Deming was giving a weekly course, the Korean War broke out.

Then, the Occupation Forces began to procure goods from Japanese enterprises and they demanded that the manufacturers deliver goods which met high standards – which meant that it was essential for the manufacturers to utilize some kind of quality control. The Japanese, for the first time, realized the widespread use of quality control in the USA, where the concept was introduced in the 1920s to ensure greater reliability of consumer durables; this had become essential because American consumers had learned to buy selectively, and in general chose to purchase those products they judged to be of higher quality. On the other hand, the dominant factor in the prewar Japanese market was procurement by the military, and the assessment of quality by the military only came under actual war conditions; hence Japanese manufacturers did not develop and use strict quality controls in connection with their production for the consumer market, as did their counterparts in the USA, where application to the production of military goods came later. In this environment, Dr. W. E. Deming's 1950 quality control seminars were well-timed. They were given serious attention by major Japanese enterprises and their engineers. Deming's ideas were first applied, with remarkably good results, to the Japanese steel and pharmaceutical industries.

The next year (1951), the Association of Japanese Scientists and Engineers began awarding the 'Deming Prize' – funds for the prize being provided from the proceeds of the sale of copies of Deming's lecture notes. In Japan, the concept of quality control has come to be associated with the name of Deming. Some Americans have questioned why the Japanese accord Deming almost total credit for introducing quality control ideas into Japan. It was essentially by some chance that Deming's lecture of quality control was accepted with enthusiasm at the time of the take-off of the postwar Japanese economy.

In 1970, Japanese engineers reached close to the level of American quality control, which they admittedly learned from American prewar precedents. Now, the pendulum has swung; during the 1970s, at the shop steward level, a 'QC movement' was started in Japan[22] in which 'bottom-to-top' suggestions for improving processes and products were encouraged by management. I entertain doubts as to whether technical suggestions made at the shop steward level are effectively adopted in the high technology situation; it is perhaps more a clever labor management measure to reduce work alienation at the workplace.

Ironically, the Americans then began to import this idea from Japan.'[23] One American observer has noted that U.S. science and engineering in the postwar period was so long dominated by military procurement that American engineers largely lost sight of the importance of continuous quality assessment and economies in manufacturing.[24]

Science and Technology in the Period of Rapid Economic Growth

The Rise of the MITI-industrial Complex – private science

In the early 50s, the slogan of 'go back to prewar living standards' was a widespread phrase among economic bureaucrats as well as the general public. The highest living standards before the war was obtained around the year 1924; since then, the Japanese economy gradually declined mainly as a result of the war effort.

Japan's production capacity was restored to its prewar level by 1955. In the same year, *Keidanren* (abbreviation of Keizai dantai rengokai, Federation of Economic Organizations, the political superstructure of Japan's big business) lobbied for the launching of a high productivity campaign. This event was so symbolically significant that some say 1955 was the starting point for the development of science and technology during the rapid economic growth period. It is, however, difficult to clearly determine any such periodization to cover all of industry, as each industrial field and its respective technology possess a different phase of development. During the latter half of the 1950s, however, the employment situation of scientists and engineers began to improve considerably; the government sector, other than MITI, was also reacting to the needs of the day as seen in the Ministry of Education (Monbusho), which in 1961 came up with a proposal to increase the number of university admissions in science and technology subjects.

During the 1950s, there was a debate in the community of industrial scientists as to the future direction for Japanese science and technology; this basically amounted to a question of to import or to become autonomous. The autonomy school, some of whom followed the Japan Communist Party official line, claimed that Japanese engineers should not be subordinate to American capital and American imported technology. This claim has certain followers and sympathizers among independentminded Japanese engineers. Hitachi was, in this respect, idealized as the best niche for engineers to work, as its board of directors had been dominated by engineering graduates. While Toyo Rayon tried to develop its own synthetic fiber technology and later became involved in a patent lawsuit filed by Dupont, the Mitsui group from the outset adopted a policy of setting up a base of imported technology. Mitsubishi Juko (Heavy Industries) chose the quickest path to bridging the technological gap by buying and learning American technology.[25] In reality, most engineers followed the realistic line of importing foreign technology, which they would later build upon and adapt. The latter was the main strategic line of MITI (Ministry of International Trade and Industry.) Ozawa has claimed that mainly during the 1950s, rather than the 1960s, imported technology played a truly significant role in modernizing Japan's industrial capacity.[26]

MITI's Function

Japanese science and technology during the rapid high growth period was by no means R&D oriented. We may be able to compare the US military-industrial complex with MITI's function in the bureaucracy-industrial complex. The major difference is, that, while the Pentagon commands an enormous budget to lavishly allocate funds for R&D efforts conducted in corporations and universities, MITI's major function is one of guiding and coordinating industries. MITI has inhouse laboratories, but most Japanese R&D is conducted by private laboratories. MITI's capable bureaucrats collect information worldwide, foresee the trends of technological development in the immediate future, advise industries in imperative tones, and if needed, organize research associations of private laboratories until they mature sufficiently so as to be able to compete with each other independently.

It may be said that another major difference in MITI's role as compared with the Pentagon is the promotion of peaceful rather than military industries. If we look closely at what MITI was actually intending however, we find that MITI's line of promotion was not necessarily pacifist. In the mid-1950s, when the Japanese industrial sector, particularly Keidanren, wanted to take advantage of post-Korean war demands from the Pentagon toward a more militaristic line, MITI also participated in the promotion of a defence industry, particularly for the area of military aircraft. MITI's motivation was, however, nothing of defending one's own country but the promotion of domestic industries, whether military or peaceful. It was the Ministry of Finance, which suppressed such a move under the stern belief that the defence industry is too precarious to support with public funds. It was thus that the Ministry of Finance prevented the formation of a Japanese military-industrial complex,[27]

MITI, however, had no consistent science and technology policy of its own, in contrast to Japanese STA and those equivalents of other advanced countries. MITI's science policy was not independent of economic policy or even subordinate to it, but more a synthesis of economic and science policy which may be more aptly termed 'industrial policy'. In other words, MITI has no tightly formulated science policy to aim for in the long term over, say, a five-year or ten-year period. It is, however, flexible enough to adapt and adjust constantly to the changing market situation[28]

If MITI had had a comprehensive science and technology policy at all, says Yasuyuki Maeda, former MITI bureaucrat and Japan's leading scholar on MITI, it was the policy of allocating foreign exchange funds to strategic industries in order to aid them in the acquisition of scientific machinery and devices needed for take off[29] Indeed, Chalmers Johnson commented on this point that 'Such a foreign exchange budget . . . lasted in one form or another until 1964, when trade liberalization was carried out. In the era of highspeed growth, control of the foreign exchange budget meant control of the entire economy. It was MITI that exercised this controlling power, and foreign currency allocations were to become its decisive tool for implementing industrial policy[30] Thus, the 1950s was MITI's heyday, and a time when it proved to be the most instrumental and powerful in bringing Japan from postwar devastation to preparing the condition for prosperity that was only realized much later.

MITI'S primary role was to protect domestic industries, and the augmentation of these protective policies through import licensing was a move against the current international trend towards liberalization. While this invited strong external criticism, the Japanese government realized that if Japan were ever to succeed in the global free market system it first had to strengthen its industrial structure. Hence, the emphasis on science and technology was part of the preparation for strengthening Japan's competitiveness and of bringing industry into line with the principle of international economic competition under the free market system.

Keichi Oshima, a representative Japanese technocrat, concluded that close coordination of technological and economic policies has been the major factor in successful technological innovation.[31]

Central Research Laboratories for Imitation

For that purpose, Japanese scientists and technologists were asked to be 'creative', rather than imitative, so that Japan could compete with others based on its own unique brand of science and technology. Toward that goal, a number of large enterprises established central research laboratories to improve the level of their technology. Thus, around 1961 there occurred a sort of boom in establishing central research laboratories (see figure 5).[32] in the industrial community. In practice, however, these laboratories were not places of creation but of copying and adapting, namely systematic imitation (in other words, stealing) of foreign (particularly American) technology.

In spite of the official 'plea for creativity', Japanese managers and engineers knew well enough that imitation is much more economical than creation. In fact, during the 50s and 60s, industrial managers sent scientists and engineers abroad, mainly to the US, with the mission of investigating new technology in the American technological community. The 'search and bring back' policy is a one hundred year old tradition, but it was particularly prevalent in the Japanese postwar business community. The actual creativity of Japanese industrial scientists was not considered as important as product marketability, capitalizing on the creativity of other nations[33]. This way of thinking prevailed even in the 70s amongst Japanese managers, and it was only in the 80s that they realized that, in order to avoid science and technology conflict with the US, they must begin to seriously rethink the old formula that imitation is cheaper than creation, a dictum that no longer seemed relevant.

Priorities of Japanese Private Science

In the prewar period, the research activity of Japanese private science enterprises was meagre. In recent times, however, their share of Japan's total research resources has increased rapidly.

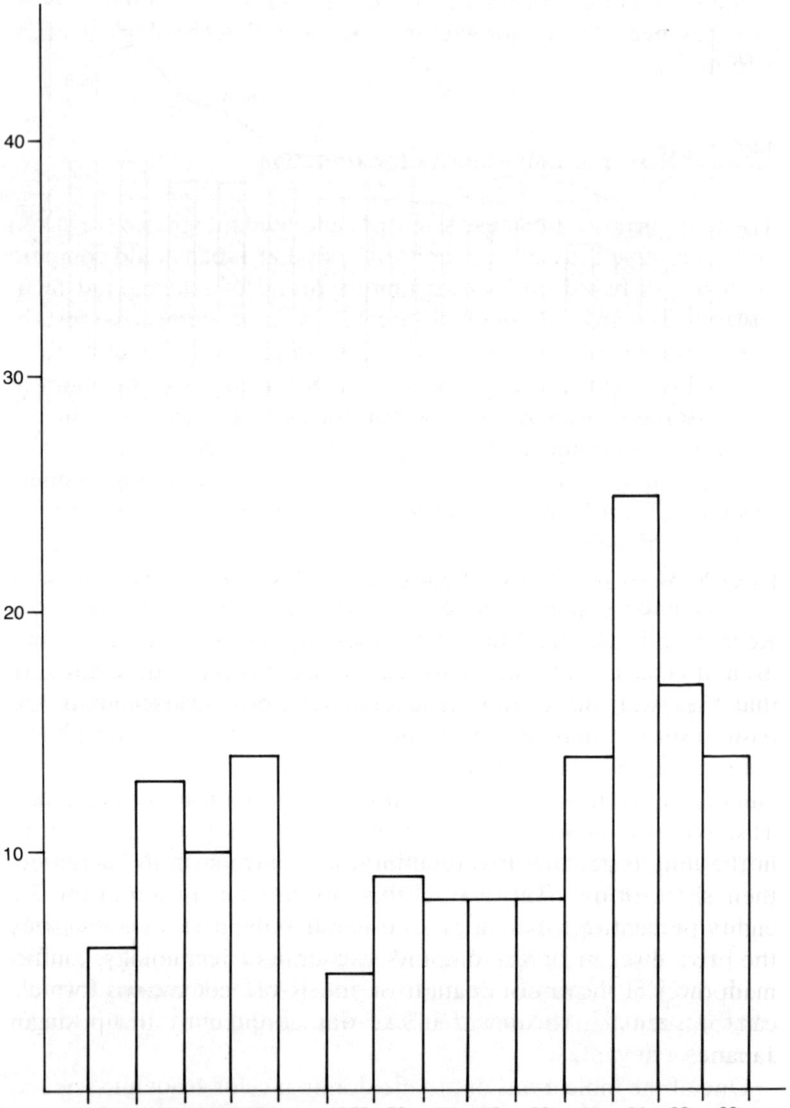

Figure 5 Number of new private central laboratories

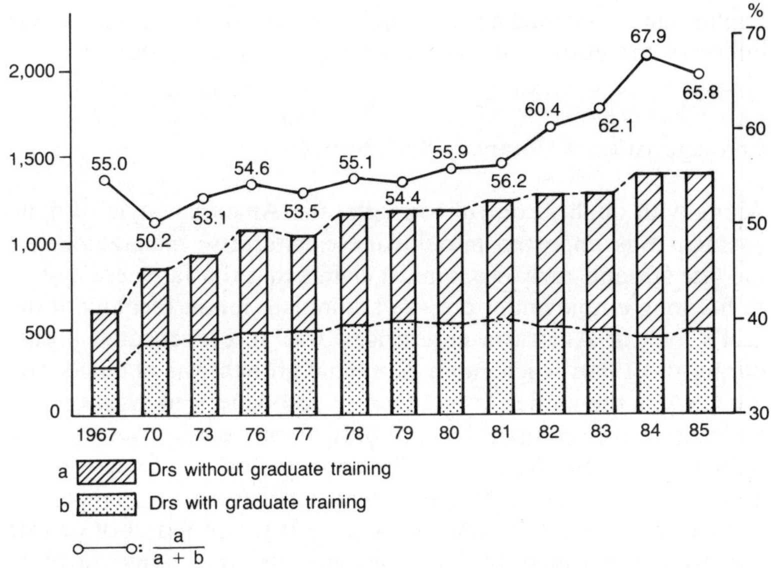

Figure 6 Number of PhDs in engineering

Recruiting great quantities of research manpower from universities and governmental laboratories around this time, they boasted that they were no longer dependent on university scientists for basic research, and claimed to be quite self-sufficient.

In the period of economic recession, private research was highly vulnerable to budget cuts and thus private laboratories were not as stable as laboratories in the universities and government. But, in the long term, the private laboratories have steadily increased their share of national research resources. Currently, more than eighty percent of total national research expenditure comes from the private sector, which employs two-thirds of national scientific manpower. Private domination of the R&D community, which can be seen in the prewar USA, was adopted by the postwar Japanese.[34]

One of the indications of private dominance of academic science can be seen by the fact that the number of Doctors of Engineering (Kogaku Hakushi), who obtained their degrees without attending graduate school but rather through submitting dissertations based on their research at private laboratories outnumber those D.Eg.s who have had regular academic training.

The disparity in numbers has increased as time has passed.[35] It implies that too strong a private influence can sometimes adversely influence the growth of regular academic training programs.

Characteristics of Company Staff Science

Along with quality control concepts, the American style of management was imported into Japanese enterprise management in the late 50s and early 60s, amidst claims that the Japanese system of life-long employment does not stimulate inhouse competition. 'Let's look at American enterprises' was an oft-repeated phrase during the 1960s when rapid economic growth was planned with the USA as the model. It is a deep-seated American management philosophy that competition not only among enterprises but also individuals within the company remains the driving force for progress. An average American scientist or engineer changes his work-place four times in his life span.[36] If the proposal of private scientists is rejected by the management, they may move to another company or form their own independent R & D company to accomplish their idea.[37] Social mobility was hailed as a means of creating challenges and a confrontation of ideas, while lifelong employment was seen as the road to stagnation until the 1970s.

During the 1980s, however, when Japan is said to have overtaken the US, because high economic growth had been largely accomplished, some Japanese people have complacently begun to conclude that all features of the Japanese system, including the life-time employment system, have contributed to the development of science and technology. We feel quite uneasy about accepting the assertion that everything is fine as far as economic growth is concerned.

If we venture to make any explanation at all, it can be made as a comparison between US and Japanese systems as follows. In the US, where there is no lifetime employment system, scientists and technologists always think in terms of moving from one place to another with the view to promoting their career and putting effort into improving their curriculum vitae by achieving individually distinct contributions, often at the expense of cooperating with others. In Japan, on the other hand, people work not as individual scientists or technologists but as a lifetime company employee discounting their 'selves' and, therefore, are able to participate in company-oriented research without reluctance. The result, if we measure from American standards, explains the lack of creativity, as creativity is often measured as individual distinction rather than group endeavor. This model is, however, the typical Japanese mode of private science, irrespective of whether ambitious Japanese scientists are happy in a company environment or not. Japanese private scientists are mostly anonymous conformists, unknown outside of the private science community. All of these factors contribute to bringing about the characteristic postwar Japanese private science.

Japanese universities and government laboratories share the same life-time employment system with private industries. This, also judging from American standards, stifles creativity and demonstrates the weakness of basic research. Out of Japanese-style private science, we can hardly expect Nobel laureates, nor major breakthroughs of technological innovation worth recording on a chronological table of the history of science and technology. One may cite the Japanese wartime effort of designing Zero Fighter planes but not the postwar Toyota car as an epoch-making dramatic achievement. Junnosuke Kishida, a well-known Japanese science journalist, has classified the targets of Japanese private science and its products into five features, namely, 1. safety, 2. economy, 3. reliability, 4. durability and 5. convenience.[38] All of these are items required in the consumer market place. To meet consumer standards does not necessarily require much creativity, nor basic science. Neither are they required to fulfill the special conditions necessary for space and military research. What is most important is the 'balance' of the abovementioned five features, which is close to the requirements of service science. Yoshiro Hoshino pointed out that Japanese steelmakers attained the balance of various imported technologies from USA, West Germany, Belgium and so on.[39] These requirements may be better arrived at through cooperative workshops than through individual competition. They are, furthermore, necessary features in a peaceful marketplace as contrasted to a war-related research effort.

In every scientific laboratory, there always exist problems of how to treat aged researchers, who can no longer keep up with the forefront of research. Under a life-employment system, this often creates obstacles to active research morale; the domination of old people suffocates youth's creativity. This is true in university and government laboratories,[40] where all researchers remain as research professionals for life, in contrast with industry where less than thirty percent of Japanese private researchers stay in laboratories all their lives.[41] In private laboratories, aged researchers are frequently transferred to the production line, where they apply their research experience to production activity, and later, return to the laboratory to apply the problematiques of the production site to research. Thus, exchange of information and personnel between the laboratory and production sites are effectively maintained. This close connection of research and production is one feature of Japanese private science.

Also it is by no means uncommon that a laboratory researcher is transferred to the sales department, where he is exposed to consumer demands. He may also bring back consumer assessment to the laboratory and production line.

In other words, Japanese private scientists are primarily company staff rather than independent professionals. To be successful company staff members, they cannot stay too long in a research position.[42] Staff move within a company, easily transferring their jobs from one place to another depending upon the firm's operational needs, while American researchers move from one company to another seeking promotion within their specialized profession.[43]

While American venture business, with its inherent flexibility, was suitable in the early stages of technological development, Japanese company-staff technology, in the mass-production phase, shows its excellence in not only product planning and system design but also in assembly, testing and quality control.[44]

All of the above contribute to strengthening the market-oriented, product-oriented characteristics of Japanese private science[45]. Combined with the mass-production system of cheap higher education, the features of present-day Japanese science and technology mentioned above might be characterized, not as the elite science of Britain or the glamorous capital-intensive science of the USA, but as 'Kentucky-fried' commercial science.[46]

Notes

1 Most renowned and widely read example is Chalmers Johnson *MITI and the Japanese Miracle* (Stanford U.P., 1982). For the opinions of

Japanese economists, see Tessa Morris-Suzuki, *A History of Japanese Economic Thought* (Routledge, 1989), chapter 5.

2 Most of the results of the U.N. University project, *Technology Transfer, Transformation and Development: The Japanese Experience* during the 1970s are addressed to a Japanese audience but its coordinator, Takeshi Hayashi in his *Gijutsu to shakai – nihon no keiken* (*Technology and society–Japanese experience*, UN university, Tokyo University Press, 1986) thoroughly discussed model export and technology transfer problems from Japan to the third world. In English, see *Japanese Experience Progress Report*, Feb. May 1978, Institute of Developing Economies, Tokyo, 1978. Most American analyses are motivated by a desire to prove that it is unnecessary to take Japan as a model.

3 I generally share Ian Inkster's opinion that the late nineteenth century Japanese experience cannot be a model for the present day 'third world'. Inkster summarizes his arguments in *Japan as a Development Model? Relative Backwardness and Technological Transfer* (Bochum 1980)

4 Even as late as 1890, the percentage of samurai graduates in each school of the Imperial University was as follows:

Engineering:	85.7
Science:	80.0
Literature:	75.0
Law:	68.3
Agriculture:	55.9
Medicine:	40.8

Source: Ikuo Amano, 'Kindai nihon ni okeru koto kyoiku to shakai ido' ('Higher education and social mobility in modern Japan'), *Kyoiku shakaigaku kenkyu* (1969), no.24, p.84.

5 Shigeru Nakayama, *Characteristics of Scientific Development in Japan* (CSIR, New Delhi, 1977) pp.48–49.

6 Sakichi Toyota's invention of a weaving machine, which was awarded domestic patents, is considered to be a kind of 'intermediate', or 'appropriate' technology to adapt to domestic conditions by converting a Western steel frame to a wooden one readily available locally.

7 Shigeru Nakayama, 'Hakurankai to tokkyo' ('Exposition and Patents'), Mitsukuni Yoshida (ed.), *Hakurankai no kenkyu* (Shibunkaku, 1986) pp.181–196.

8 Shigeru Nakayama 'The Transplantation of Modern Science to Japan', *Center for Studies in Higher Education, the University of California, occasional paper*, no.23 (1982).

9 Yoshiro Hoshino, *Gijutsu kakushin*, (Iwanami).

10 Tetu Hiroshige et. al. *Nihon kagaku gijutushi taikei, tsushi 5* (*Com-*

prehensive history of Japanese science and technology, outline history no.5.)

11 Chikayoshi Kamatani, *Gijutsu taikoku hyakunen no kei* (One-hundred year plan for the great technological nation, Heibonsha, 1988).

12 Shigeru Nakayama, 'Science – a National Endeavor', *Bulletin of the Visiting Scholars Association*, (Harvard-Yenching Institute, Japan branch, 1974) vol.4, pp.13–19.

13 *MITI* p.20.

14 Keichi Oshima, 'Research and Development and Economic Growth in Japan' in B.R.Williams ed. *Science and Technology in Economic Growth* (Macmillan, 1973) p.313.

15 Junnosuke Kishida 'Sengo gijutsu no taishitsu o kimeta mono' ('Determinants of the characteristics of post-war technology'), in Nakayama (ed.) *Nihon no gijutsuryoku* pp.45–49.

16 Occupation documents ESS-ST, 'Reports on Japanese Laboratories'

17 Personal correspondence from Dr. Donald J. Pletsch, ex scientist officer of the Division, February 1989.

18 Yoshiro Hoshino 'Sengo gijutsushi no jidai kubun (Periodization of postwar history of technology)' in Nakayama (ed.) *Nihon no gijutsuryoku* pp.75–86.

19 Hoshino, 'Sengo' pp.84–85. Also Tomijiro Murata, 'Sengo Nihon gijutsu no 30nen' ('Thirty years of postwar Japanese technology'), *Keizai hyoron*, vol.24, no.8 (1975), pp.46–50.

20 Hiroshi Ichikawa, 'GHQ kagaku gijutsu ka no seikaku to katsudo nitsuite' ('On Policy and Activities of GHQ-ST'), *Osaka Shidai Ronshu*, no.54, p.5 (1987).

21 Dr. Bowen C. Dees, private correspondence, in January 19, 1989.

22 J.M.Juran, Jo Yamaguchi and Kaoru Ishikawa, 'QC no kokusaika jidai' ('International Quality Control'), *Hinshitsu Kanri* vol.21, no.3, p.184ff. (1970).

23 *MITI*, p.216.

24 Nathan Rosenburg, personal communication, 1982.

25 Junnosuke Kawamura, 'Sengo nihon no gijutsu donyu no keiken 1950 nendai no jirei' ('Japanese experience of importation of technology in postwar period – Case studies in the 1950s') in Kokumin keizai kenkyu kyokai (ed.), *Kokumin keizai* (*National economy*) no.147. (1983) pp.25–53.

26 Terutomo Ozawa, *Japan's Technological Challenge to the West, 1950–1974*, (MIT, 1974) p.37.

27 Hideo Otake ed. *Nihon seiji no soten* (*Controversies in Japanese politics*, 1984) chapter 1.

28 In this sense, the Ministry of Public Works (Kobusho) in the 1870s and 80s, had a more independent science policy and hence many of their projects turned out to be failures.

29 Talk to Science and Society Forum, 1987.

30 MITI p. 25.

31 Keichi Oshima, p.320. who explained the function of MITI officials in pp. 320–321.

32 Shigeru Nakayama 'Sengo no mondaiten (Postwar problems)' in Isao

Sugimoto (ed.) *Kagakushi* (Yamakawa, 1957) chapter 13, pp.478–481.

33 Matthews, p.176.

34 Given the different criteria used by each country, it is virtually impossible to make sound international comparisons of budget structure and scientific manpower. It is, however, meaningful to discuss a national trend as long as one employs the same relative indices over time.

35 'Rikokei Daigakuin' pp.45–46.

36 Statistics were compiled from *American Men and Women of Sciences*.

37 J.M.Juran, Jo Yamaguchi and Kaoru Ishikawa, 'QC no kokusaika jidai' ('International Quality Control'), *Hinshitsu Kanri* vol.21, no.3, p.185 (1970).

38 Kishida, 'Sengo gijutsu no taishitsu' pp.54–56.

39 Hoshino, 'Sengo gijutsushi' p.95.

40 Largely for this reason, industrialized scientists do not regard public and university researchers very highly. Sometimes, exceptionally creative university researchers are consulted by industrial scientists but this is not the general case and for public laboratories this is much less so.

41 Science and Technology agency ed. *Kagaku gijutsu hakusho* (*White paper of science and technology*, 1988) p.36.

42 'Rikokei Daigakuin' p.97.

43 Gary Saxonhouse, 'Technological Progress and R & D Systems' *U.S.Japan Science and Technology Exchange* p.49.

44 Masanori Moritani, quoted in Sigurdson p.68.

45 While sociological aspects of academic science have been studied recently by a number of sociologists of science, the mechanism of industrialized science has been hardly explored, because, unlike the former, the community of the latter is not quite open to outsider investigators and also because the systems (reward system and other infrastructure) are all different from one company to another. The statements here are generalisations, a result of student (Tokyo University, Science Club, 1987) interview investigation of three types of scientists (private, public and university researchers) and my own experience of having joined the Human Resource Projects of *Shigen chosakai* (Institute of Resource Investigation) of the Science and Technology Agency in 1984–85. While university and public researchers are outspoken and critical of their own laboratories, private researchers have a tendency to tell us successful aspects only, with the public relations of company to which they belong in mind. The latter do not like their personal name to be referred to by student investigators.

46 The name 'Kentucky fried' science is derived from an Australian TV program, 'Kentucky Fried Medicine' (1989), in which the privatization trend of Australian socialized medicine was critically assessed. In my application of the expression, however, I do imply not only a negative, namely 'privatization' of science, but rather a positive sense of the word, namely, that Japanese postwar science and scientists

have been cheaply mass-produced and well quality-controlled, resulting in science which is oriented toward the consumer market only and is, therefore, 'science for profit'.

6 Weakness of National Projects–Public Science

Japanese public science has mainly been conducted by STA. In this chapter we contrast the behavior of STA with MITI by analysing some of STA's major national projects such as nuclear and space research.

Absence of Big National Projects

In 1954 on the same day as the Bikini incident mentioned above in chapter 2, diet member Yasuhiro Nakasone put forward what was then a huge budget for research and development of atomic energy in accordance with 1953 President Eisenhower's plea for 'atoms for peace', project D. For that mission technocrats proposed the creation of the Science and Technology Agency (hereafter abbreviated as STA). In spite of the *Minka* group's protestations that STA would promote the way toward militarization, as exemplified by nuclear armament, it was established in 1956.

It can be generally said that in the structure of R&D funding in Japan, priority has always been given to the industrial sectors under MITI's protection, while the status of big national projects of the STA in such fields as nuclear energy and space remained relatively low throughout the postwar period.

STA has complained in its *White Paper on Science and Technology* (annual publication) that the Japanese government does not allocate as much to science and technology as other advanced countries do, thus indirectly criticising the Ministry of Finance, the body in charge of allocating the national budget. At budget allocation meetings within the government, the Ministry of Finance has persistently responded that the major reason for Japan's relatively small science and technology budget is because Japan does not conduct military research and emphasizes only R&D for economic development; if one excludes military research, the R&D budgets of other nations is similar to Japan's level of government expenditure on R&D[1] This is supported by the calculation made by the National Science Board of the USA which compares the ratio of total R&D spending to GNP as compared with that of civilian R&D. The ratio of Japanese civilian R&D to GNP exceeded that of the U.S., U.K. and France, in the 1960's and was only inferior to West Germany[2]

Throughout history, the biggest concern of a nation state has been the defense of its own existence. It is not surprising, therefore, that government expenditure for science and technology has been allocated predominantly to military affairs, though in modern states welfare aspects have also come to be emphasized. In this regard, postwar Japan is quite exceptional in limiting the defence research budget to less than one percent of total national expenditure on R&D. Furthermore, defence scientists have often been denied admission into academic societies, where an anti-military ethos usually prevails.

In the 1980s, however, the Japanese military research budget gradually increased its percentage from 2 to 5 within the nation's R&D effort[3] with the help of outside pressure from the US government, the Reagan Administration, which stipulated that Japan must duly share the defence budget of the Western allies.

Japanese Disproval of Effective R&D Public Policy

In the midst of the postSputnik science and technology boom, the first half of the 1960s saw a rapid increase in public spending for R&D in all major industrialized countries. A science policy was needed to allocate the science budget. A rationale for research investment was also needed for public understanding of science. In order to meet these demands, there emerged a group of people who seriously attempted to calculate quantitatively the cost of R&D allocation vis-a-vis the benefit of economic growth on a national scale. Certain suggestions and proposals followed the research findings, one of which recommended that R&D allocation should at least be 3 percent of GNP. However, the major conclusion of the study was that investment in basic science will cause technological innovation and eventually lead to economic growth[4].

In the United States and the Soviet Union, R&D allocations for science and technology have been a part of competition for national prestige or for military purposes, while in European countries and Japan economic considerations have been the major criteria for such allocations. Calculations, at first, are made so as to overestimate the benefit that investment for R&D contributes towards economic growth. The research planning bureaucrats were often charged that their approach is akin to 'throwing a sprat to catch a whale', and that behind the proposals for increasing R&D allocations lay their desire to expand their own influence. In this kind of calculation, there is always an element of arbitrariness and confusion about the basic assumptions. However, this situation was later amended to a more modest estimate.[5]

The correlation between economic growth and R&D investment remains a popular belief. But when it comes to an international comparison, the differences in basic concepts according to the countries under investigation differs greatly. It becomes apparent by examining Japanese funding levels, that the Japanese case disproves the validity of the assumption, that there is a positive relationship between investment and economic growth. High growth was realized in the sixties when R&D investment was low, and as R&D investment was steadily increased, growth has slowed. Japanese high growth, then, was attributable to other factors entirely. The discovery of such negative correlations in various settings frustrated the whole science policymaking community and led to policy change and drastic budget cuts in the USA, England and France in the latter part of the 1960s. OECD, the stronghold of science policy makers, in *Conditions for Success in Technological Innovation* (Paris, 1971) concluded that 'there is no observed correlation between the proportion of national resources devoted to R&D and the rate of growth in productivity.[6]'

Since then and throughout the 1970s and 1980s, all of the advanced nations took lessons from the Japanese experience implicitly,[7] and gradually reduced the public sector share of total R&D. USA's non-defence public obligation was reduced from 47.2% in 1965 to 28.9% in 1987, though its defence R&D was

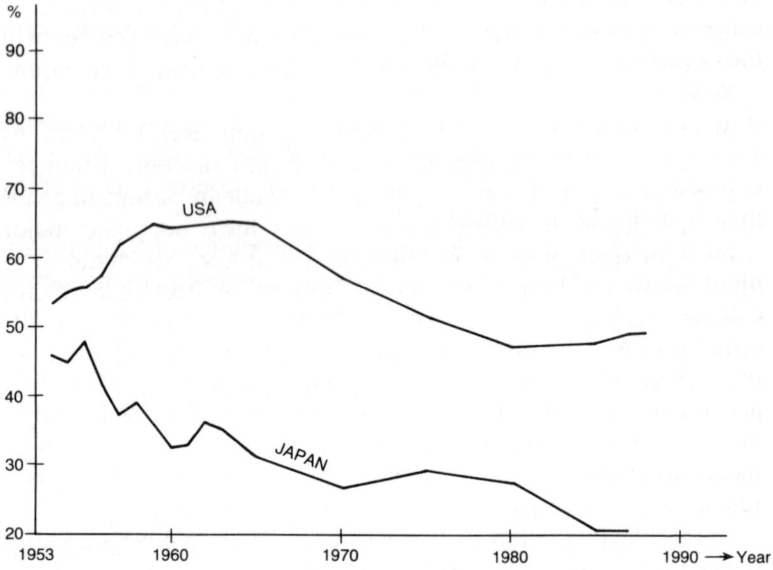

Figure 7 Public sector obligation of total national R & D in percentage
Sources: *Kagaku Gijitsu Hakusho* (Science and Technology White Paper), p. 341 and p. 376 of 1965 section and p. 124 of 1988 edition.

not significantly reduced. The French non-defence commitment was reduced from 62.3% in 1965 to 32.0% in 1986, though its defence R&D increased.[8] (see fig. 7)

Unlike England or France, Japan has had no grand R&D plan like SST (supersonic transport), yet despite this Japan has been able to realize rapid economic growth.[9] This poses the problem of disputing the belief of technocrats in Western advanced countries concerning their theory of R&D investment. It invalidated the belief in the eventual accountability of academic science, the idea being entertained by postwar science policy makers since Vannevar Bush's *Science, the endless frontier* published immediately after World War II. The attitude of the Third World science policy makers may be summarized as follows:

Wisely staying out of involvement in costly basic science development, which does not guarantee economic return to society, Japan borrowed the basic ideas and research results from the United States and other countries, using them only in an economically viable line of development, without expending large amounts of money. Japan has, as a result, been able to reconstruct its economy and attain high economic growth. We too have to learn the secret of their success and copy it.[10]

With a sense of self-criticism and sarcasm, such was the new style of evaluation of Western science policy makers when viewing Japanese behaviour in the 1970s, critics who, at one time, once enthusiastically claimed the necessity of economic growth by means of R&D investment.

Such being the case of the apparent bankruptcy of R&D investment theory and the benefits of research, simple equations such as calculating the percentage of GNP which is the most appropriate optimal value for R&D investments are no longer relevant. Moreover, R&D investment for defence science is never commercially accounted for. What the research budget is for, for what purpose money is to be used is a much more pertinent question. Furthermore, apart from research *per se*, there is the problem of the social relevance of research and development. Thus, the philosophy of sciencebudgeting policy, which so far had been established during the period of unrestricted development of science and technology in the period of high economic growth in the early sixties, became the object of fundamental reexamination upon entering the third phase of 'the age of questioning.'

We may also note the emergence of the contrast between the behavior of the militaryindustrial complex and MITIindustrial complex or between American and Japanese models, which became clearly noticeable only in the 1980s.

Sectorial Structure of Science Funding Compared

Drawing on the four sector representation of science funding, we shall consider a few indices crucial in science policy. In order to gain a comparative picture of the style of science policy.

The public-private ratio is represented by budget-obligations, while academic science is represented by expenditures, as it is obvious that the financial reserves deriving from universities is minimal. Though intended as a purely visual device without serious quantitative rigour, governmental non-defence funding roughly corresponds to service science and defence funding to (incorporated) public science in the foursector approach.

There are two obvious patterns in this representation. One is to find two conspicuously different types, the Japan-West-Germany type and the USA-France-UK type, or more simply the Japanese and American models. The other is to be found by comparing structures of the 60s and 80s where it is seen that all countries move from public funding to private, or move from the American to the Japanese model. Now let's consider these two patterns in more detail.

In the postwar period, we may be able to single out two distinctive science policy approaches, the American and the Japanese. While the former is led by the state-military according to strategic considerations, the latter is led by the industrial sector with governmental development assistance and protection. The former is promoted only under the Cold War political climate. While most major industrial countries followed the American model, the Japanese enjoyed comparative advantage under the American defence umbrella to adhere to its own model. Only when the cold war climate is cleared up, does a Japanese model become attractive to the other advanced countries; as that occurs, the Japanese will loose the comparatively advantageous position of avoiding defence expenditures.

Pseudocivilian Technology

In the USA, it has always been accepted that big science associated with defence and space may not be directly beneficial to the economy, but that spinoffs from such research become the seeds of future technological innovation; such thinking still prevails among people who are the promoters and supporters of the SDI (Strategic Defense Initiative) plan. However, after entering into the 1980s especially when the comparison with Japan has become more apparent, there have appeared some doubts about this belief among certain opinion leaders. Nathan Rosenberg is one of them. According to him, American scientists and engineers have been supported throughout the postwar period by federal expenditure; therefore, they have cared little for profits in pursuing their research and development.[11] In other words, they have been too academic. Hence, when compared with Japan, which has pursued R&D policy for economic benefits placing primacy on industrial production, the American weakness from an economic point of view is obvious. National aid for science and technology may

Figure 8 (1–5) Four-sectorial representation of the structure of national science policy

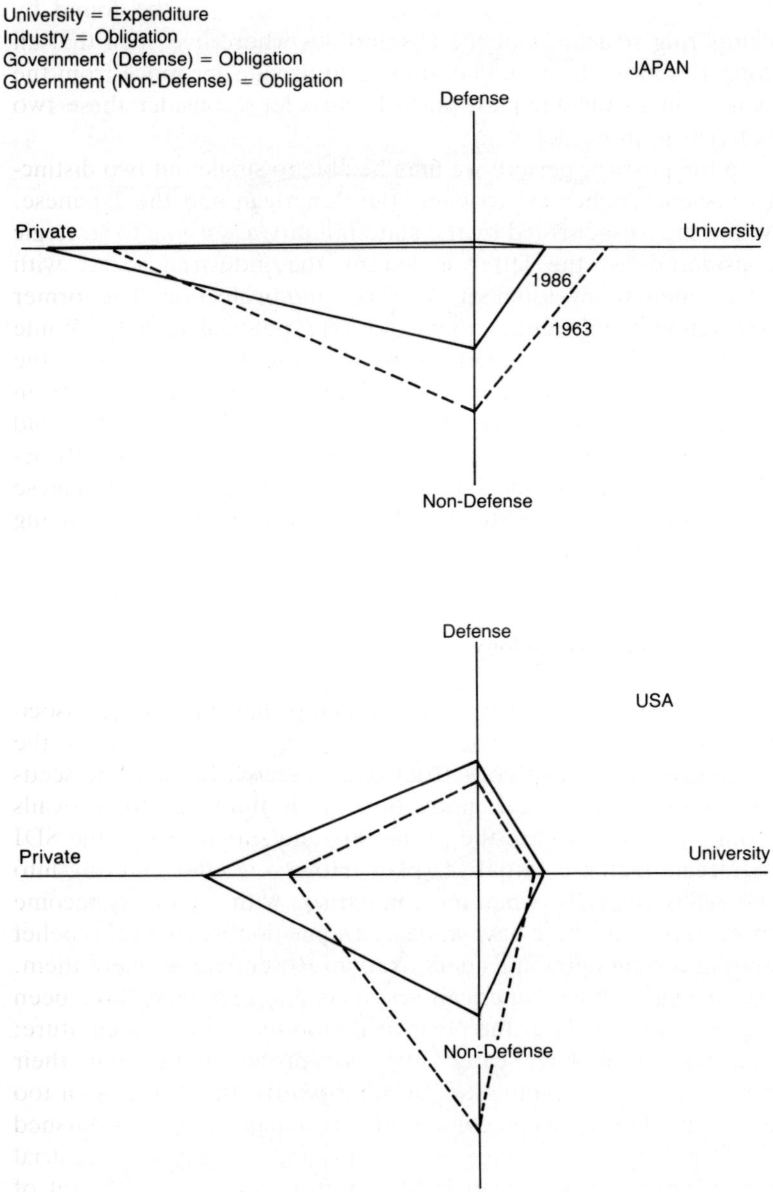

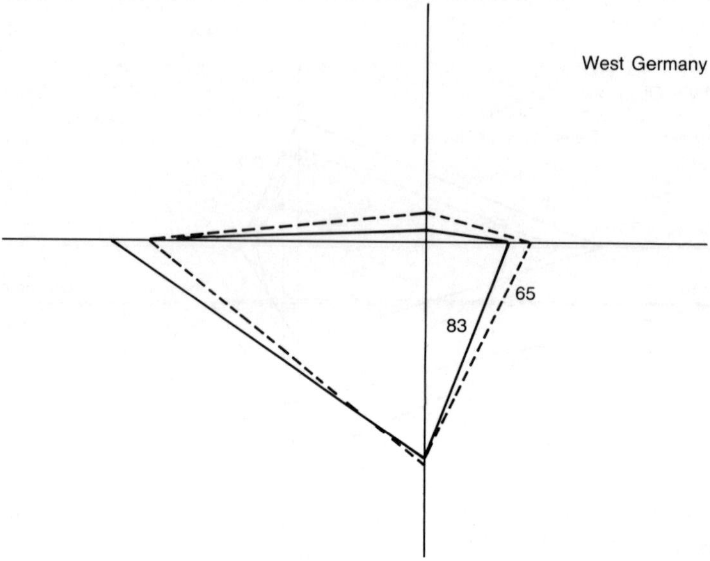
West Germany
65
83

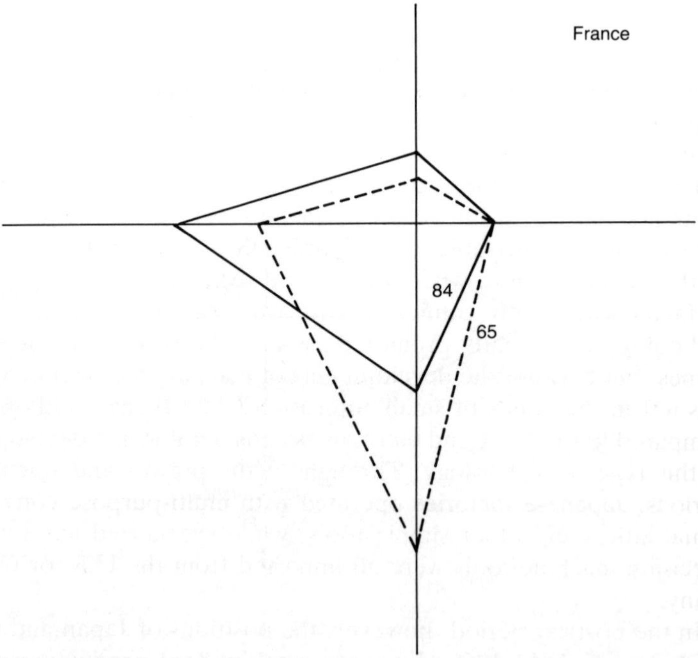
France
84
65

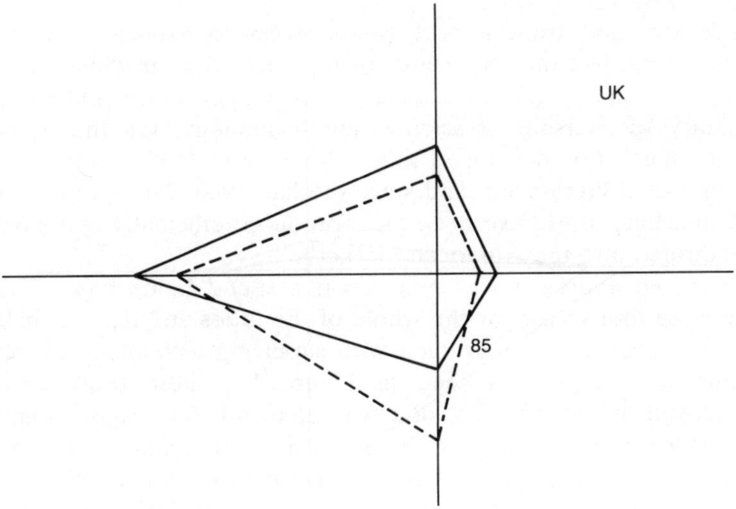

Source: STA, *Kagaku Gijitsu Hakusho* (Science and Technology White Paper, 1988), pp. 400–416.

provide stimulation in the short term, but inexpedient in the long term, since it spoils the development of science and technology in alliance with the market economy.

This sort of example is quite clearly seen in the machine tools technology which is the basis of manufacturing technology or, indeed, the indispensable basic component of industry in general. In the prewar period, when science and technology development in Japan was closely connected with war requirements, military technology was greatly promoted, as in the case of Zero fighter planes, but because the developments of machine tool technology was left in the hands of small enterprises[12] like Ikegai Tekkosho, compared to the USA and Germany a considerable gap developed in this type of technology. Throughout the prewar and wartime periods, Japanese factories operated with multi-purpose conventional lathes and other simple tools, while specialized automatic precision machine-tools were all imported from the U.S. or Germany.

In the postwar period, however, the positions of Japan and the USA reversed. In 1966, Japanese machine-tool production and trade switched from import value excess to exports excess[13]. Despite the fact that NC (numerically controlled) machine tools originated in the US, they have not developed much under the military sponsorship of science and technology. On the other hand, Japan first developed this technology in 1955 on the basis of imported literature,[14] and since the late 1960, Japan's lowcost NC machine tools have been mass produced efficiently and have penetrated into the American market[15].

Such an approach was built upon a sound home base. The Japanese tool sector for the whole of the fifties and the first half of the sixties was preoccupied with supplying technology to the domestic market[16] As seen in Figure 1[17], these tools were employed to satisfy the domestic demand for major home appliances (TV, washing machines, refrigerators, and so on), the markets of which came to be saturated around 1970. Following the satisfaction of domestic demand, the Japanese industries sought an international market.

In the 1950s and 1960s, microelectronic technology developed dramatically as a result of spinoffs from basic research related to military and space. In the 1970s when this technology was commercialized for the production of consumer goods, the highly sophisticated and advanced militarytype technology, developed to meet the specialized conditions of high pressure, high temperature and so on, was not useful for everyday life. Thus, these two directions of technological development, for military and consumer purposes, have become increasingly independent.

Microelectronic technology is rather an exceptional case of a successful spinoff, at least in its early development stage. Most of the closed public science of defence and space have not created spinoffs into the civilian sector for technical or organisational reasons. Technically, a technology which meets the requirements of war and space is not directly applicable to everyday life. More importantly, the institutional infrastructure in which closed public technology is promoted does not fit well into the open market environment. Aant Elzinga calls this kind of technology 'pseudo-civilian technology'. Nuclear technology is a case in point.

Atoms for War and Peace

Rabi's Base Laboratory Idea – An Occupation Episode

During the Occupation period, research in nuclear science was strictly prohibited. As the Cold War developed, it was likely that in some quarter of the US army an opinion emerged to permit such research by Japanese atomic scientists, who were frustrated by the prohibition and entertained apprehensions of not being able to catch up with the forefront of research. The following is an episode I have recently discovered from topsecret documents of the Occupation Forces.

In November and December 1948, I.I. Rabi, an American Nobel laureate in physics, came to Japan as a member of the US Science Advisory Mission sent by the (American) National Academy of Science. On December 18, after visiting Japanese research laboratories and interviewing leading members of the Japanese scientific community, he discussed with Dr. H.C. Kelly, who was a former physicist member of the MIT Radiation Laboratory under Rabi during the war, financial help for Japanese scientists, who were suffering from postwar poverty. 'For example,' said Rabi, 'in Professor Sagane's (an experimental nuclear physicist) research at the University of Tokyo, the addition of $ 2,000 or $ 3,000 to the budget would enable them to get their equipment into effective operation very quickly and they would produce results of value . . . The sums involved are small, but the effects would be large, both in production of scientific results and in production of a surplus of trained personnel in experimental physics which could go effectively into industrial research and development.'[18] The last part of this quotation echoes the New Deal line which Kelly tried to promote, namely economic recovery by means of sceintific research.

Accordingly, Rabi wanted to appropriate some of the US military funds given to American universities, which he knew were inefficiently used because of the shortage of personnel in the USA. He did 'not prefer to classify projects, but rather to provide unrestricted support for basic research leading to the development'. On December 10, Kelly sent copies of the above memoranda to his superior, General Marquat, the head of the ESS, with his own strong recommendation to implement Rabi's suggestion.

On the same day, Rabi sent a letter directly to the Occupation

Army to suggest 'The Use of Japanese Research Facilities as an Advanced Base in the Event of Acute Emergency in the Far East.' Citing the successful precedent of the advanced base laboratory in Britain from his own wartime experience, he stated 'it is recommended that we permit, encourage and aid the Japanese to establish one or more laboratories, particularly physics and telecommunications, . . . to furnish such a base for our use in case of emergency.' Aware of the situation in which the Cold War mentality was advancing at the time, Rabi took advantage of it to persuade the military establishment to support his idea. His true hidden intention seems to have been aimed at the encouragement of academic scientists in Japan, for whom he felt sympathy as a fellow member of the scientific community. Attached to the proposal were detailed data and evaluations on eighteen leading laboratories in Japan selected out of thirty-two laboratories visited by Rabi's mission with preference for the application of research which would contribute to the economic recovery of Japan. Of these eighteen, eight laboratories were listed as worthy of receiving active assistance from the United States in order that they might better qualify to serve as base laboratories to be used by the United States in the event of an emergency. Rabi did not forget to add, in accordance with Kelly's line, that these laboratories would have an important favorable effect on the Japanese economy in a direction beneficial to American interests.

As a step in the implementation of Rabi's idea, Kelly sent from the ESS a check sheet dated 12 January 1949 to each department of the Occupation Force for approval and comment. Among the responses of agreement from the departments, one worth quoting is by G-4 (General Staff Section 4, in charge of logistics) dated February 3. G-4 argued that a nonmilitary research and experimental laboratory system in Japan would actually conduct military work and advised that 'Unless surrounded with the utmost secrecy and its purpose adequately concealed, the project might be used by powers unfriendly to the US to prove that the Americans were furthering military research in Japan or were even allowing Japan to conduct its own military research in preparation for war. Hence, the proposed action should be first referred to the Department of Army at home. Furthermore, as it is conceivable that US use of the laboratory facilities would involve advanced electronic and/or nuclear physics projects, it would seem advisable to coordinate it with the Research and Development Board, National Military

Establishment.' It was in this context that the Occupation considered permitting research by Japanese in such hitherto forbidden fields as nuclear physics and radar.

Upon receiving these responses, Kelly sent out a further check list on March 17 to emphasize his own position that 'ESS recommends that the program if adopted be directed solely toward economic purposes, though ESS does not object to G-4's suggestion to inform the Department of Army of this proposal.' On March 19, SCAP (Supreme Commander for the Allied Powers) sent a telegram on this proposal to the Department of Army.

I have not succeeded in finding the response of the Department of Army, nor any further sources. It does not seem that during the Occupation permission to conduct research on the atomic bomb was granted to the Japanese. However, it is an interesting episode which helps to elucidate the interrelationship of interests among the scientific community (both American and Japanese), civilian NewDealers and the military establishment.[19]

Public Science of Nuclear Projects[20]

In order to meet the needs of the Korean War, the GHQ issued a memorandum to revise the prohibition of Japanese weapon production to make it possible with the permission of the GHQ in March 1952, [21] but research on the atomic bomb was never permitted during the Occupation period. Right after the Peace Treaty in 1952, a committee for weapon research was created within the Keidanren. Although they discussed guided missiles, there is no reference to atomic bomb production.[22] It was only after Eisenhower's plea for atoms for peace in 1953, that the Japanese scientific as well as business community took nuclear energy seriously.

Commercial Failure of the National Projects

The first notable example of big public science in the postwar period is found in the STA's unsuccessful venture into autonomously promoted nuclear energy production.

During the Occupation, Japan was strictly forbidden to conduct nuclear research. Despite the feelings of frustration that prevailed within the academic and industrial scientific sectors as Japan was left far behind the international frontier of nuclear science, scientists were not well prepared when in accordance with the 'atoms for peace' plea of US President Eisenhower in December 1953 a budget for nuclear reactor construction (Nakasone budget) passed the Diet in March 1954. Various elements of an infrastructure were hastily organized in academic, public as well as industrial sectors, to promote nuclear research and development, which culminated in the establishment of the Japan Atomic Energy Commission and the STA in 1956, the office to promote a big national project. At the time, the national project dealt with nuclear R&D only. Those who participated in the national project envisaged eventual nuclear energy autarchy by an autonomous Japanese science and technology sector.

There was another line of development, made possible through the importation of foreign technology, promoted by the MITI-industrial complex, consisting of MITI, the nuclear industry (reactor and equipment manufacturers, Toshiba, Hitachi and Mitsubishi, etc.) and public utilities. The MITI-industrial complex followed the conventional postwar catch up procedure of relying on foreign technology, starting with the importation of a Light Water Reactor (LWR). It proved remarkably successful, as the nuclear energy industry became commercialy profitable and Japan is now (in 1989) the fourth largest nuclear country, with thirty percent of total electric power consumption generated through nuclear means (see Fig. 9).

The STA and its associated public corporations, such as the Japan Atomic Energy Research Institute and the Power Reactor and Nuclear Fuel Development Corporation, have from the outset adhered to the self-appointed goal of developing a domestic Fast Breeder Reactor (FBR) and later an Advanced Thermal Reactor (ATR) and Boiling Water Reactor (BWR). Their plan has been promoted with a considerable portion of the budget and hence has attracted a high level of visibility in the mass media. It has so far, however, been dogged by delays and, at this stage, can be deemed a failure economically, as is true of all Western attempts. In the 1980s, when STA sought to transfer their demonstration reactor to the private sector, the latter showed resistance and reluctance.

From the point of view of the MITI-industrial complex, the national projects of the STA and its associates can be criticized

Figure 9 Nuclear energy's yearly international comparison

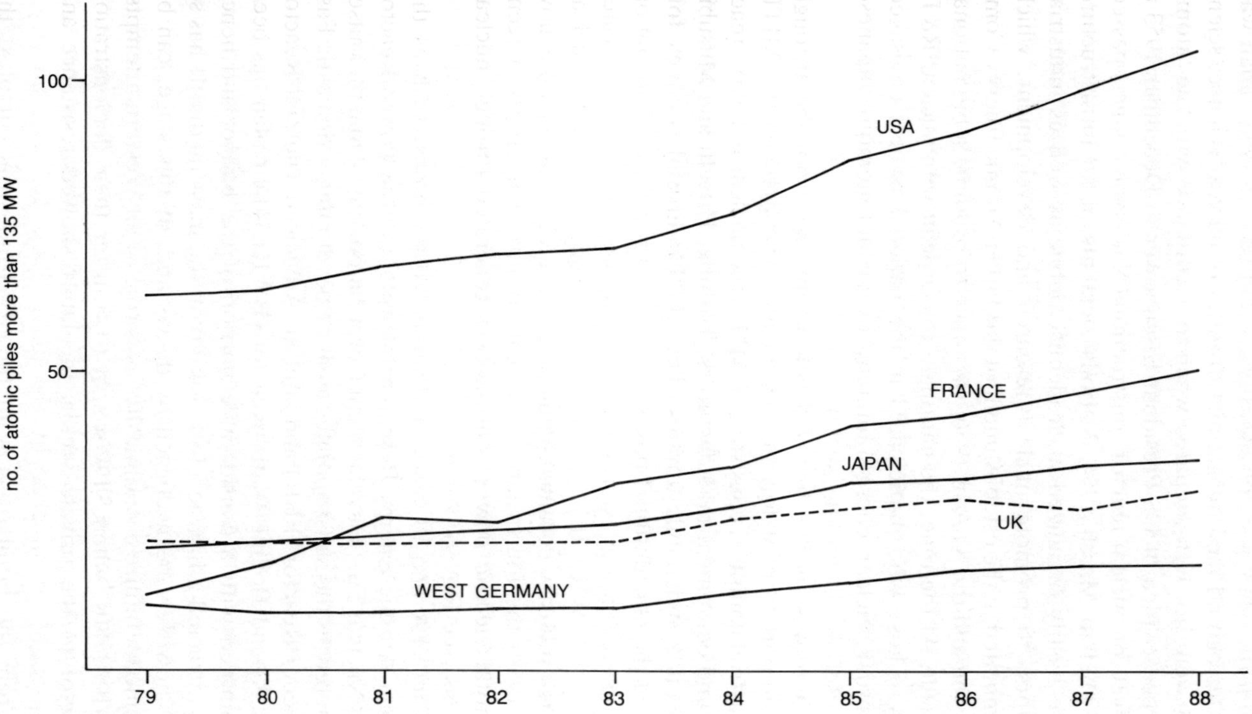

USSR: no data.
Source: *Genshiryoku Hakusho* (Nuclear Energy White Paper), 1989, pp. 336–337.

as follows: The postwar technology transfer to Japan has been successfully carried out by first importing foreign technology and then step by step gaining autonomy from a previous position of dependence. Unlike other areas, such as microelectronics, the STA in the area of nuclear power did not follow the conventional pattern used by MITI but from the beginning tried to develop its own autonomous national project with large government expenditures. Despite large expenditures, STA failed while MITI's approach to their LWR line succeded. However, unlike West Germany and France, the Japanese nuclear industry is still based on imported licences, which prohibits the possibility of export. If MITI's line of development were followed with adequate subsidy, Japan might have reached the stage of technology export by now, which is the customary goal in other Japanese industrial areas. In other words, STA and its national project has been nothing but an obstacle to its own purely commercial development.

The Japanese Atomic Energy Commission could have attempted to coordinate the above two different lines of development, but AEC, in fact proved to be a mere consulting body, and had no power of enforcement.

Raison d'être of National Projects

However, national project should not be assessed solely in terms of their economic development and profitability. The raison d'etre of a national project exists in the area where commercial accountability is not possible. In pseudo-civilian technology, such as the nuclear reactor, assessment can be concomitant with essentially non-commercial goals, such as defense. In the USA and elsewhere, the cost of the essential part of nuclear technology R&D was recovered with the strategic defence value of nuclear bombs. Postwar Japanese nuclear technology has significantly lacked this core part that conflicts with the postwar pacifist national sentiment. Although the international community retains a lurking suspicion in regard to the ready convertibility of Japanese nuclear reactors into bombs, military use has never been mentioned by those Japanese officials in charge of nuclear plans, for fear that they might be attacked by the mass media as being unconstitutional. In the absence of military motivation, Japanese public science of nuclear projects offer as compensation academic interest and national prestige, while economic development competes with the interests of the MITI-industrial complex. The academic sector generally has little influence upon the public sector except for those engineering professors who sit on the committee meetings of STA-related nuclear projects with their academic viewpoint and pride, and these professors must, therefore, support autonomous R & D rather than mere importation and imitation of foreign made reactor technology.

The status of STA bureaucrats must be evaluated and assessed in proportion to the budget they secure. They found their raison d'etre in promoting big public science as comparable to the big national science projects of major Western nations, a notable example of which is the American space program. Without having established close connections with the private industrial sector, the motivation tends to be set for its own sake and the assessment criteria of nuclear R&D under the STA umbrella remains 'academic' (in the sense that prestige and reputation are sought rather than practical applications), if not militarily strategic.

There is a 'civil strategic' rationale to the STA line of development. Though STA, like other independent agencies attached to the prime minister's office, was in part controlled by MITI through its detaches at its outset (48 MITI detaches out of a total 76 detaches), there is a rumour that some sections of MITI were opposed to the creation of STA, as their basic philosophy and ideology were considerably different and even conflicting. While MITI is, as its title suggests, the promoter of international trade, an autarchic insulationist policy can be perceived in the stance of STA. Though STA denies this the predecessor of STA may be traced back to the wartime 'Gijutsuin' (Office of Technology) which was the stronghold of prewar and wartime technocrats.

The Office of Technology bureaucrats were most concerned by the scarcity of resources in Japan. In order to maintain a strong modern nation, they entertained thoughts of an autarchic economy of natural resources, petroleum energy sources, iron and aluminum ores and so on, to prepare for the time of crisis when resources cannot be imported. In fact, the rationale of starting the Pacific War has been solely attributed by resource bureaucrats to the USA-initiated resource embargo, which prewar Japan desperately tried to break. They had planned a Pan-East-Asian recource autarchy by exploiting iron, petroleum and tin etc. from the surrounding areas. In the postwar period, however, they found it impossible to exploit the natural resources of surrounding countries; instead, they had to find an alternative way by relying on science and technology. In place of resource autarchy they hoped to create a science and technology autarchy. For them, science and technology information is nothing but resources.

The newly-created STA shared the basic philosophy of the old Office of Technology and thus STA aimed for selfsufficiency of energy (and science and technology) resources by creating autonomous nuclear technology, at the time when abundant Middle Eastern petroleum was not available. They believed that this was a national project (kokusaku), which should not be relegated to the unstable private sector. Civil strategic resources, such as energy and foodstuffs, are 'public goods' to be nationalized within the control of the the public sector. In the postwar period, natural resources may be replaced by science and technology resources in terms of priority, which must be promoted and kept within the public sector because of the strategic importance of science and technology. It was the autarchic insulationist school, rather than the international school, which encouraged the development of indigenous scientific technology.

In the postwar period, there has been repeated controversial dissent between the internationalist and isolationist camps on the matter of strategic resources; as time passed, it became increasingly apparent that Japan would successfully stay out of war in the future. (Such a long lasting peace was not conceivable for the prewar generation). MITI's strategy has proved to be successful and the private sector has risen, the insulationist's nationalistic plea has faded out, only to be revived in times of oil crises.

Nuclear fusion project

Like other major Western countries, the inception of Japanese nuclear fusion research was inspired by the celebrated words of an Indian physicist, Homi Jehangir Bhabha, at the first UN international congress on the peaceful use of atomic power in August 1955: 'I venture to predict that a method will be found for liberating fusion energy in a controlled manner within the next two decades.' It remained academic until the end of the 1960s, when the STA participated in a big way.

In the early academic period, the spirit of postwar laboratory democracy still prevailed as the physics community made decisions about such big resource intensive projects. It was termed a 'democratic centralized system', somewhat similar to the organizational principles of the Japan Communist Party. All of the opinions of the community were gathered and debated at the Nuclear Fusion Special Committee of the Japan Science Council.

Two plans emerged. The A-plan sought to promote basic academic research at the university level. The B-plan aimed at a research and development plan for a middlesized nuclear fusion device to generate and enclose high temperature plasma. The latter was supposed to be conducted under the STA umbrella. Although the two projects are not incompatible but conceivably able to be developed simultaneously, the decision was made to give priority to the A-plan, for which purpose the Plasma Research Institute was granted affiliation to Nagoya University in 1961. Academic fusion research was also undertaken at the Universities of Kyoto (Heliotron) and Osaka (Torus and Laser). The B-plan was revived in 1966 and the fusion R&D plan entered into the second phase in which STA and its associate Japan Atomic Energy Research Institute (JAERI) have come to play the major role. A transition was made from academic physics to public science.

The Japan Atomic Energy Commission, which is the ultimate decision maker for public nuclear planning, has published a long-term plan every five years since its inception. Whereas it had been devoted mostly to fission programs, a fusion plan appeared for the first time in the third long-term plan published in 1967. This implies that nuclear fusion research and development was officially recognized and a big budget was granted.

According to this plan, a fusion reactor would take roughly five years from design to operation and another five years for testing and collection of data. Out of the data thus accumulated, the following steps repeat a cycle of the design of the larger reactor, test and data-collection, until the final goal of a commercializable reactor is attained within the first two decades of the twenty-first century. Following this planning, the first public fusion reactor, JT-60 (JAERI Tokamak 60), was planned in 1974, the design phase began in 1978 and was completed in 1985.

It seems, however, that the peak of enthusiasm for the fusion project was in the 1970s. Perhaps because of the technical and financial difficulty in attaining the projected time table, the timetable for commercial feasibility has disappeared from public reports since 1982. Bhabha's two decades of lead time has never been shortened but rather tends to be lengthened. Frustration among those seriously engaged in such a project is apparent. Many of them seek consolation in the possibility of alternative application of their accumulated knowledge.

The Case of the Space Program

The Japanese space project started in 1958 as academic science at a laboratory affiliated to the University of Tokyo as a part of the International Geophysical Year project of launching a rocket for scientific observations. Following the Sputnik incident in 1957, the government came to prepare public support for the space program. Since 1960, the budget increase for the Tokyo University project under the Ministry of Education auspices was accelerated year by year until 1967. In 1964, the Institute of Space and Astronautical Science was set up, affiliated to the University of Tokyo. Academic scientists from the outset sought autonomous research and development; the use of satellites was limited to scientific research. They achieved steady progress in geophysical surveys in the 1970s. Astronomical observations since 1979 have resulted in the development of an internationally recognized area of X-ray astronomy.

STA started to promote the public science of space in 1963. In spite of considerable advances achieved by the Tokyo University academic space project over the public one of STA, STA obtained hegemony over the Japanese space program in the late 1960s, far exceeding the Ministry of Education budget since 1968. This change-over may be partly attributed to the fact that the Ministry of Education and academic science suffered from student protests between 1968 and 1972, though university-affiliated laboratories were less affected. More important, however, the national policy of space R&D was established in this period.

Since the early 1960s, the demand of the public sector for utility satellites, such as those for weather forecasting, broadcast and communications, put pressure on STA to adapt imported technology, in order to make it utilizable as early as possible. Accordingly in 1968, a US-Japan agreement was concluded to permit the transfer of American space technology to Japan, on the condition that it should be limited to peaceful uses and should not be transferred to other countries. The postwar pattern of technology transfer, namely building up its own technology on the basis of imported technology, was belatedly applied here also because of the classified nature of American space technology.

Thus, in the early 1970s, public space programs depended on American rockets and satellites, while academic space science still pursued an autonomous line of development. Autonomous utility satellite programs have been taken up again in the beginning of the 1980s. With the development of Japanese microelectronic technology, Japan is expected to undertake major space programs in the 1990s. Unlike the nuclear program where the commercialization of electricity generation is the goal, the Japanese space program was confined to the non-military public sector. Hence MITI remained in a subsidiary position whilst the major line of development was drawn by STA.

What is public science for?

There are three kinds of Japanese public laboratories;

1. Service science laboratories, such as agriculture, health care, prevention of pollution and natural calamity, etc., where the private sector stays out because of the lack of a market mechanism.
2. National big project laboratories under the STA umbrella, such as nuclear power and space, which do not yield profit in the immediate future but is to be supported according to a long-range national policy.
3. Industrial science laboratories under the MITI umbrella. Each category has around 5,000 scientific manpower, totalling 15,000.[23]

The first category is not evaluated in terms of academic or private science but service science. Its activities must be supported even if a dramatic success is not expected.

The decline of public science is not limited to STA-related activities. The MITI-related inhouse public science, such as that of the Agency of Industrial Science and Technology, was once located on a nationwide research strategy chart in between academic and private sciences. It was expected to conduct seminal basic research for later private applications. However, its activities have been increasingly overshadowed by the rise of private science. The private sector, competent enough to conduct seminal research by themselves, tends to disregard the role of public science. Due to its structural mechanism and the lack of profit incentive, public science cannot compete with private science in the pursuit of science for profit and economic development. Public science is now experiencing a crisis of clear motivations and goals.

It is, however, the indispensable role of public science (including those of public corporation laboratories) to assess private science and its products for public use. A wide extension of this role to non-profit-oriented service science, such as the R&D of robotic helpers for disabled people, is considered to be a unique feature of publicly supported science.

Notes

1 *Gakujutsu geppo*, vol.23, p.177 (1970).
2 *Science Indicators 1978* (National Science Board, Washington DC, 1979) pp.140, 144.
3 *Kagaku gijutsu hakusho* (Science and technology white paper, 1980 and 1989).
4 It was the official claim of Japanese STA. STA, *Kagaku gijutsu hakusho* (Science and Technology White Paper, 1962) pp.59–61.
5 Such efforts are well summarized in Bela Gold (ed.), *Research, Technological Change, and Economic Analysis* (Lexington Books, 1977). 'To help answer this question, the literature of the past twenty years has been analyzed from the standpoint of its applicability in assessing the actual effects of actual innovations rather than the theoretical effects of hypothetical innovations under artificial conditions.' (p.1).
6 OECD, *Conditions for Success in Technological Innovation* (Paris, 1971), p.20. See also OECD, *The Research System* (Paris, 1974), vol.3.'
7 Although science policy bureaucrats never explicitly doubted the ultimate validity of research spending and the value of science policy study, I have heard their complaints at various occasions for science policy studies in the early 70s that Japan presented a damaging example to their profession.
8 STA (ed.), *Kagaku gijutsu hakusho* (1988), p.124.
9 George C. Eads and Richard R. Nelson, 'Japanese High Technology Policy: What Lessons for the United States?', Hugh Patrick ed. *Japan's High Technology Industries* (1986) p.243.
10 Ron Matthews, p.182ff. P.B.Stone, *Japan Surges Ahead-Japan's Economic Rebirth* (London, 1969) p.138.

11 Personal communication, 1982.
12 *Nihon kagaku gijutsu shi taikei, 18, kikai gijutsu* (Compendium of the History of Japanese Science and Technology, no.18, Mechanical Engineering, Daiichi Hoki, 1966), chapter 14, section 2.
13 Ron Matthews, 'Technological Dynamism in India and Japan: The Case of Machine Tool Manufacture', E. Baark and A. Jamison eds. *Technological Development in China, India and Japan* pp.158–159.
14 *Nihon kagaku gijutsu shi taikei, kikai gijutsu*, pp.566–567.
15 Matthews, p.172. Also James Coates and Michael Kilian, *Heavy Losses* (Penguin Books, 1986) pp.362–263.
16 Matthews, p.164.
17 Changes in the ownership of major home appliances, *White Paper*, 1986.
18 All the quotations in this section are derived from microfilm of the ESS-ST in Occupation Documents preserved at the Diet Library, Tokyo, that were copied from the sources preserved at National Recording Centre, Suitland, Maryland, USA.
19 Shigeru Nakayama, 'Rabi, kichi kenkyusho keikaku jiken' ('Base laboratories planning'), *Kagaku asahi*, September 1988, pp.83–87.
20 This section is largely based on the preliminary work of and discussions with Hitoshi Yoshioka, which will eventually appear as a volume of the Toyota project on 'Science, Technology and Society in Postwar Japan'.
21 *Shiryo sengo nijunenshi* (Twenty years history of postwar documents, H. Arisawa and S. Inaba (eds.), Nihon hyoron, 1966) vol.2, p.165.
22 *Ibid*. pp.166–168.
23 Reona Ezaki and Tetsuji Nishikawa (eds.), *Kagaku kakumei, Nihon ikinokori no senryaku* (Scientific Revolutions: Strategy for the Survival of Japan, Daiamondo, 1983) pp.231–232.

7 Grassroots Revolt – Possibility of Service Science

SECTION 1. Turning Point from High Growth

In the late sixties and early seventies, the unprecedented high economic growth of the preceding period came to an end. This trend was initiated by students and citizen revolts against the establishment. In this section, we examine the effect of citizen movements on the change-over in public and private sectors in their science and technology policy.

When we conduct research on the development of science policy up to the high economic growth period, it will be necessary to start with Cabinet resolutions and follow the process of the policy implementation that flowed from top to bottom.[1] In the period of reflection and structural change during the 1970s, however, the situation changed completely and policymaking power entered the hands of those participating in citizens activities. Growing public concern generated in the citizen sector made its impact on the public sector to such an extent that the Environment Agency in 1971 was established as an outcome of this process. Mass media and environmental movement groups heralded and represented the concern and opinion of citizens.

Student Revolt and Criticism of Science

The university stands at the interface between science and society. Most citizens still believe that science is pursued within an ivory tower, even though the postwar phenomena of mass higher education has turned the ivory tower into a more accessible popular entity. On the university campus, high science and mass education have had to coexist. It was way back in the early nineteenth century when the coexistence of education and research was developed to its highest degree in the German universities under W. von Humboldt's ideal. Since the war, the university has raised its ivory tower as high as it possibly can, at the cost of sacrificing links with society, in order to maintain supremacy among the scientific communities. On the other hand, in order to respond to the popular demand for higher education, the university has been greatly enlargened at the expense of its traditional elitist approach. While the concepts of quality and quantity are not necessarily compatible, universities have had a problematic existence, being pulled in two different directions: highly specialized science and ever-expanding mass education.

In 1968, institutions facing such problems throughout the world were hit by student revolts. Students were no longer the avant-garde, select few but a powerful group amounting to two million in the case of Japan, able to bring about social revolution all by themselves. Their impact on the scientific community and its institutions was formidable.

Japanese-language literature on university problems rapidly increased in 1968 and reached its peak in the following year, as shown in figure 10.[2] If we examine it closely, we find that the

Figure 10 Publications on university problems

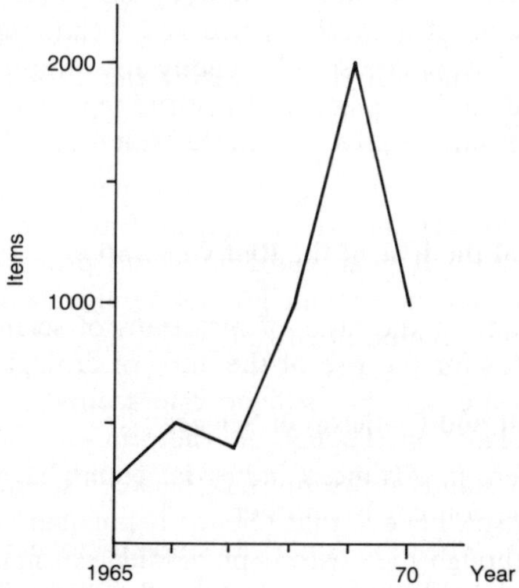

Source: *Kagaku to Shakai no Gendaishi* (Contemporary History of Science and Society), p. 84.

literature from 1968 to 1969 mostly deals with student movements, discussed not only by academics but also by politicians, businessmen and other sorts of people. Literature in 1970 mainly dealt with the topic of university reforms. Academics responded to the problems raised by the student movement in 1968–69 with various plans for university reform.

Throughout the history of student movements, two directions have been observable: an internal direction, institutional reform of universities and an external direction, socio-political reform. In the case of postwar Japan, the demand for laboratory democracy right after the War belongs to the former category, while the latter culminated in the movement against the U.S.-Japan Security Treaty in 1960.

Student movements of the late 60s differ markedly from the above-cited precedents in their unique stance of anti or critical science. The leaders of the movement postulated that science and technology had become a tool of the oppressive technocratic regime, under which the university scientist exists merely as a rank and file employee. While in modern history representatives of other cultures often criticise the coersive nature of modern science and technology, in the late-60s movement it was young professionals and students who had been destined to become the main promoters of scientific activity, who fundamentally questioned the value of modern science and technology. Their ideological quest was, of course, not readily answerable and could not be responded to with piecemeal institutional reforms. Hence, it became more introspective and more enduring.

Explanation of the Rise of the Riot Generation

It has been one of the tasks of historians of science to examine possible causes for the rise of the anti- or critical science generation in the late 60s, but still no one seems to have given an adequate explanation. Though the rhetoric was shared by certain student leaders in advanced industrial countries, the conditions vary from one country to another.

In the 1950s, when the American student movements were dead and a silent generation was created, an education commentator wrote that perhaps student movements were found only in developing countries, where frustrated elite students in higher education became leaders hoping to overthrow the existing regime. In highly industrialized societies, the general public was so highly educated that students were unable to raise their voices to be heard by adults. The 1960s proved this not to be the case. In the USA, the anti-Vietnam war and anti-draft campaign must have contributed to the rise of the new riot generation. The French May Revolution involved labour unions and other citizens. These are all missing in the Japanese case. Able to avoid committing itself to the war effort, Japan in 1968 enjoyed an unprecedented, high level of economic growth, a growth of more than ten percent each year. As the economy grew, students had little to worry about with regard to future employment. An international comparison based on questionnaires given to students in 1969–70 reveals that while the lack of employment possibilities was attributed as one of the major causes of the student movement in the Western world (France 91%, Italy 88%, UK 47%, USA 31%), Japanese students who considered it so numbered only 7%.[3]

The dissenting student generation of the late 60s coincides with the generation of the postwar baby-boom. When those born right after the war reached college-age, they found higher education crowded and facilities inadequate. During the Occupation period, Japanese public higher education was expanded by order of the Occupation Forces, along with private universities too in the early 1960s. In terms of quantity, universities are more open to those who want higher education than European students. But Japanese students are similar to the French and Italians in attributing the causes of the student riots to the poor quality of the university system.[4]

Ideological Cause

Whereas criticism[5] of the structure of a nation's scientific community varies from one country to another, ideological causes are more commonly shared by students and the younger generation in postindustrial societies.

Prior to 1968, postwar science and technology had a positive image. Economic recovery by means of science and technology was a strategic formula deeply embedded in the minds of postwar Japanese. The formula proved to be successful in the 1960s, elevating the Japanese economy to a level higher than its prewar standard. Postwar Japan was perhaps the most optimistic country in promoting science and technology. The optimism obscured the problem of pollution in the early 60s. High school students were indoctrinated in and inspired by this formula to choose their career in the science and technology professions.

After 1968, when they entered the university, they were shocked to find that science and technology was no longer called for in a heavily polluted society. The former catchword of 'development' had become a negative symbol. Chemistry, once very popular because of the success of polymer chemistry, became one of the most unpopular disciplines for students.

The problem of 'two cultures' existed even in technologically optimistic Japanese society. There has been, however, little confrontation of the two cultures on campus. Unlike the Anglo-Saxon university structure, Japanese universities have been divided and separated into different schools from the outset in the nineteenth century, thus missing opportunities for intellectual confrontation. Only the campus crisis of the late 60s provided opportunities for the professors and also students of different schools and departments to get together to discuss university problems, since both were temporarily unable to pursue research in their speciality because of the campus disturbances. Some academics recall this as the only merit of the campus disorder.

The writings of Jacques Ellul, Herbert Marcuse, Joergen Habermas and Theodore Roszak were translated into Japanese during this period and widely read amongst students. The ideological criticism of science has certainly liberated many undergraduates from the widespread popular faith in scientism, the belief that science and technology should be absolutely right and therefore promoted. It provided those students with the ability to look at scientific activity in relative terms.[6]

Structural Criticism

While young undergraduates were not fully committed to the scientific profession and their criticism of science tended to be ideological and abstract, young professional scientists and graduate students tried to focus their critical attention on the more concrete institutional environment of science.

Many of them revived the demand for laboratory democracy

which was strong in the early postwar period but which had been largely hollowed out during the high growth period. Physics departments had been by and large democratized during the earlier wave of democratization, but this time the demand for change became fierce in democratically backward areas such as medical schools and departments of forestry, where the conservative academic hierarchy had been tightly maintained.

In earlier times science was at the cutting edge of rationality and objectivity, fighting against the conservative feudalistic system and the legacy of wartime militarism, but this time it was science itself that was criticized. While formerly science was the liberator, the critics now found that science had become incorporated into the technocratic system during the high growth period. What was wrong, then, with science?

Out of their debates, a critical science school has emerged. It is not based on an overall anti-science feeling shared by literary critics and philosophers, but its aim is the analysis of the structure of contemporary science, especially that of privatized, incorporated technocratic science, and to pinpoint the pathology of the contemporary scientific establishment. Those who have or had practiced science and who were thus familiar with the character and behaviour of the scientific community became some of its most active proponents.[7]

Academic Response

The response of the academic sector to criticism made by students, ideological or structural, was limited to matters of institutional reform, laboratory democracy, and the pursuit of efficiency and rationalization of systems in education and research.

The minority of faculty, mostly capable and efficient, who had been frustrated by the idle inactivity of conservative bureaucratic practices took advantage of the pressure of student power and proposed various university reforms.[8] Though fundamentalist students were not enthusiastic for a mere institutional reform, a number of reform plans appeared one after the other written by various groups of people, not only academics but politicians, old graduates and citizens. One estimate shows that 429 reform plans were produced by 94 universities and 104 plans by non-academic sectors.[9]

All the plans sought laboratory democracy; and most of them urged student participation in university affairs. Some radical ones proposed abolishing graduation certificates in order to eliminate the so-called 'examination hell' and academic cliques (gakubatsu). As to the research system, some proposed the abolition of academic hierarchy by treating professors and assistants in the same manner. Some claimed that the tenure system should be replaced by appointment renewal every five years in order to more actively promote academic achievement. A careful look at these documents discloses the fact that most of them gave up the old Humboldtian synthesis of teaching and research and proposed a two-tier system clearly separating the two traditional functions of universities.

Few of these plans were implemented. As student power diminished, nobody paid further serious attention to them. Throughout the history of universities we scarcely find a radical university reform by university members themselves, unless it is related to the empowering of the majority. In fact, most of the reforms are proposed by the avant-garde minority and the silent majority neglect them unless an external force compels them to follow.

Since major university reforms were dependent on the bureaucratic control of the Ministry of Education, reforms were blocked. For example, a certain university permitted student participation in the election of a new dean but the elected dean was denied appointment by the Ministry. Thus, the bureaucratic control over Japanese universities was too strong to allow any radical reform on the part of universities, comparable to the French May Revolution or American affirmative action.

On one occasion during the student revolts, even the hard core of the Ministry of Education was influenced by the radical ideology of the time, but their response was slow and meager. After the big wave of student radicalism had passed, the Ministry of Education made a small number of reforms to soften the bureaucratic rigidity of the system. The length of residence for the doctoral candidate, which used to be set at five years, now has became flexible according to the student's ability and discipline. It has now become possible to request that faculties of other universities be allowed to participate in the assessment of a doctoral dissertation, etc. These are largely matters to be left to the academic sector rather than questions of bureaucratic control.[10]

In placing the Japanese effort of university reform in the world context, we may say the following. The French system of higher education had been inflexible under the centralized public sector dominated system since Napoleonic time and so sluggish in adapting itself to the needs of the changing times that various kinds of negative effects accumulated, finally inviting revolution and major reform. In contrast, because of the significant absence of a central power of control, American higher education has been constantly and flexibly adapting itself to the demands of society; affirmative action being one of the major issues to be adopted in the context of the '60s and '70s. Japanese universities are placed in between the French and American in terms of flexibility, but due to the centralized power of the Ministry of Education, the reform remained piecemeal and minor.[11]

New Model Universities

Radical student bodies adopted a 'scrap and build' strategy of university reform. It is more time consuming and energy consuming to scrap an old university rather than build a new university. In the beginning of the revolt period, radical students were preoccupied with scrapping the old university and its system, but after realizing the difficulty, many turned to building various alternative schools. Most well-known and long-lasting was Jun Ui's ecology school, 'Han-daigaku koza (Anti-university seminars)', which lasted nearly two decades. But most of them, lacking official recognition and sponsorship, died out within a few years. Instead, local governments and mass media have started adult education programs called 'culture centres' which bring together citizens belonging to the more affluent sections of society.

The Ministry of Education's effort was focused on the creation of Tsukuba University, which was planned in 1969 and opened in 1974. It aimed for modern rationality as opposed to traditional bureaucratic inefficiency and formality. It was modelled after new model universities abroad in the 1960s, such as Sussex and the University of California at San Diego. It was, in reality, no more than a mere location move of an old university,(the Tokyo Bunrika University), whose faculty was split over the adoption of the new model. The humanities and social sciences faculties generally opposed the new model, while natural scientists, who badly needed newly equipped laboratories, supported it.

Furthermore, a new model which was framed to fit the high economic growth period of the early 60s did not match the spirit of the period of revolts. In spite of the Ministry's atypical support in financial as well as legal terms, the new model university did not become a model to be followed by others. The Ministry itself was troubled by opposition from the academic sector and never attempted to create similar ones. Instead, responding to the need for scientific workers, the Ministry has attempted to create new types of technical colleges from time to time, but hardly on a scale comparable to Tsukuba University. Because of financial constraints in the public sector in the 1980s, no new noteworthy model has been proposed.

From Student to Citizen Movement

Chronologically speaking, the Japanese campus struggle started on January 29, 1968, when the students of the medical school of the University of Tokyo entered into a strike for an indefinite period. On July 2, a University-wide Student Struggle Committee (Zenkyoto) was organized. In the following year, student unrest and violence spread to most colleges and universities and campuses were occupied by students. However, the energy of the student struggle was in its later phase diverted into the sectarian strife between JCP and other factions.

Students' energies were directed into the university administrative core and they were successful in paralyzing most of its functions. Chanting slogans for perpetual confrontation and university reforms, they organized blockades of university facilities to make the campus a liberated zone. Some of them, stimulated by the simultaneously occurring French student movement, went out of campus to enlarge their 'liberated area' in the city and demonstrated their commitment to the citizenry, but few obtained the cooperation of labor unions as the French did when they urged a strike by millions of the labour force. Apart from minor young radical groups, the majority of labor unions did not show sympathy towards student movements, as the latter did not interest the former with economic issues. Therefore, the Japanese student movement failed to shake the political establishment as the French counterpart did in inducing major educational reform. While the 1960 movement was focused on an attack on parliament and bringing about the resignation of the Prime Minister, the 1969 revolt concentrated on internal campus affairs without threatening government bodies and without impressing citizens. During the 1960 movement, many faculty members were friendly to and cooperated with the student anti-establishment struggle, and some were able to remain indifferent and apolitical. In 1969 all faculty were involved and academic science was severely damaged. This invited emergency measures for administrative control of universities.

The political leaders considered measures to bring the campus struggles to an end but they showed neither sympathy nor interest in university reform. On August 17, 1969, the Diet adopted the *Daigaku un'ei rinji sochiho* (Law of Extraordinary Measure for University Administration), a temporary law for university management to impose penalties on universities experiencing unrest. According to the law, universities which permitted prolonged student blockades were to be deprived of their university status by the government. Thus, fearing closure, the troubled schools and colleges invited the riot police to suppress the movement. Faculties were, up until this time, reluctant to allow the police force onto campus for fear that it might encourage further violence. But since academics were fearful of enforced cancellation of their university charters, academic decision-makers easily turned to the police force for the resolution of campus struggles. Subsequently, nationwide campus unrest was nearly terminated by the end of 1969.

Those student activists and participants in the struggle were now driven from the campus. Frustrated youths joined in citizen anti-pollution and civil rights movements, resulting in a change of focus for the grassroots revolt from the campus to pollution sites.

Citizen Revolts

Science and technology did not have much of a direct influence on the change of people's lifestyle until high growth started to produce consumer durables. On the positive side, people enjoyed using the products of high economic growth to transform their lifestyles. For Japanese lifestyle, three symbolic home consumer durables in the first wave around 1956 were: the refrigerator, TV set and washing machine, and, in the second wave around 1960:

the piano, air-conditioner and private car. On the negative side, local people opposed the establishment of nuclear research laboratories in their neighbourhood, and suffered from inner (food) and outer (air and water) pollutions.

Up until the 1960s, however, the 'Grassroots sector' or 'citizenry' vis-a-vis scientific and technological activities had not been clearly identifiable. The year 1970 marked the mass media's active participation in an antipollution campaign, followed by a sudden upsurge of publications on pollution, which reached a peak in 1971–72 (fig. 11).[12] Though citizen concern is hardly measurable in a quantitative manner, the number of publications is certainly an indication of citizen response as these books are mostly sold on a commercial basis on the open market.

Air and water pollution, of course, go back to prewar times, but in those days human rights were not well established and city smog was even affectionately boasted of as a symbol of heavy industrialization. Furthermore, most pollution in prewar times was confined within factories and industrial cites, often being treated as occupational diseases. During the postwar period, the

Figure 11 Publications on pollution (Japanese and translated studies)

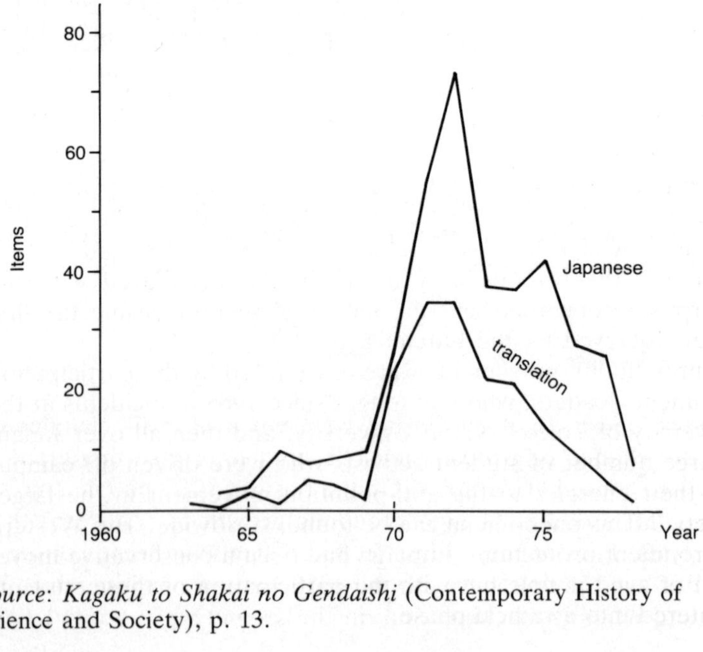

Source: *Kagaku to Shakai no Gendaishi* (Contemporary History of Science and Society), p. 13.

extent of pollution rapidly grew during the time of high economic growth and consequently spilled out to the environment where ordinary citizens lived. These are not new problems for the scientific community but the social cost of high growth. Unionists in the scientific community often claimed that there was nothing wrong with science but that the entire responsibility lay with the capitalist establishment, which abused science.

The new postwar problems which did not exist in prewar time were, however, rooted in the postwar development of science and technology, for which the scientific community has at least partial responsibility. These were: the physical pollution of radioactivity which accompanied the development of nuclear power; and the chemical pollution of pesticides, drugs, food additives, synthetic detergents, and industrial waste. When the scientific community is diffident in carrying out its social responsibility, citizens revolt.

Anti-pollution Movement

1970 was the crucial year when the media anti-pollution campaign became so strong that it was able to sway public opinion. Its origin can perhaps be traced to the USA. In that year public opinion polls regarding environmental issues frequently appeared in the press and consequently public awareness of pollution grew to an unprecedented level.

During the 1960s, Japanese high economic growth was attained by fulfilling domestic demands for consumer durables. Then, around 1970, the supply of major home appliances reached saturation point. Consumers who had their basic needs satisfied then came to realize the social cost of their needs in the form of pollution, which the mass production system had created[14].

At first the responsibility for pollution was placed with the enterprises concerned and the industrial sector became the first target, followed by the bureaucracy.

Antipollution movements were invigorated by the participation of students leaders, who had fought since 1968 in incidents at the University of Tokyo, Nihon University, and then all over Japan. A large number of student activists who were driven off campus lent their energies to the anti-pollution movement in the larger society.[15] This phenomena can be found worldwide. The Western environment protection campaign had been a conservative movement of aged people but with the participation of these students it entered into a radical phase.

In Japan, the rampant Minamata disease (mercury poisoning) had been associated with industrial discharge as early as 1956 but government and industrial sectors together tried to prevent the investigation of the true cause and thereby neglected the plight of victims.

In 1971, however, what came to be known as the four major litigations against pollution were won by the people and behind these was the power of vitalized citizen movements at work.

Public Sector Response – Technology Assessment

During the period from 1966 to 1968, there emerged a difference in awareness of pollution problems amongst bureaucrats. Whereas the Ministry of Health and Welfare and Ministry of Agriculture and Forestry started to undertake an examination of pollution problems as part of their official duties, MITI and STA neglected them, still adhering to their high growth policies. In 1968, however, the latter agencies were compelled to adjust themselves to new circumstances. They then found a temporary shelter to hide in with the new concept of 'technology assessment' that had just been brought from the USA.

A preconception of 'technology assessment' is 'technology forecasting'. Erich Jantsch, a consultant to OECD, published *Technology forecast in perspective* in 1967, which was translated into Japanese in the following year. In the preface by T. Kristensen, the Secretary General of OECD, some misgivings on the impact of technology are noticeable. Jantsch's concept was also criticized by Robert Jungk as pointing to a society like a military organization, which may pave the way for an authoritarian, totalitarian technocracy.[16]

In contrast, in the preface to its Japanese translation, Hideo Fujita, the managing director of the Japan Management Centre, stated that 'We have now come to recognize how advantageous it is to lead others in technological innovation. Hence, we have to preemptively find the future of technology in order to overcome the forthcoming crisis of (capital and trade) liberalization.'[17] He still clings to the 'catching up' mentality common amongst Japanese technocrats in the high economic growth period.

They share an optimistic view of the future. Prewar Japan had a major war in every decade when economic and technological development was fuelled by military imperatives. Since 1934 the living standard had been declining to fall to its lowest point at the time of surrender. In contrast, the Japanese after World War II experienced continuous economic growth, despite minor recessions. They now came to optimistically believe in the possibility of the extrapolation of past trends into the future. For this purpose, they employed technological forecasting in order to develop promising technologies systematically and efficiently in an optimal way. They never thought of controlling and restricting technological development. Instead, they maintained a view of bringing forth a future society which would be adaptable to future technology. In other words, they were stern believers in the technological fix.

Around that time, the US produced a concept of 'technology assessment' to predict the impact of negative factors on the environment and society in advance before new technologies were actually developed. In 1967, the Technology Assessment Act was drafted by the US Congress.

The Japanese seem, at first, to have confused technology assessment with technology forecasting. It was officially 'introduced' in 1969. In the fall of that year, the futurists of 'The Society of Science, Technology and Economy', a club of technocrats, sent a special investigation team to study American industry forecasting. On November 15, they reported that they were impressed by the antiwar demonstration in Washington DC, while they were shocked by the fact that the lunar landing of Apollo 12 (second to Apollo 11 which landed on July 20) did not cause any sensation among the American people. At the same time, they were surprised that people were discussing the 'technology assessment' concept, wherever they visited. As they had so far modelled themselves after USA in formulating science and technology policy, they conferred as to how to import this new concept immediately after their return. They had, to date, imported positive things from the advanced Western countries. If negative factors appear, they also emerge early in the advanced countries, and so do countermeasures. Thus, pollution and its countermeasures are all exogenous concepts rather than endogenous. (This attitude was clearly seen in the case of recombinant-DNA guidelines around 1978.)

As they were influential with respect to public policy, their opinions were adopted in the official guideline report of *Science and Technology Policy for 1970s* by the Council of Science and Technology, in which technology assessment, life science and soft science were emphasized. Following this report, in 1971 STA started case study projects on technology assessment for agricultural chemicals (pesticides and chemical fertilizers), high-rise buildups and computer aided instruction, and in turn MITI conducted studies for nuclear energy and steelwork.

STA and MITI projected a number of technology assessment projects during the 1970s but the concept became unfashionable and faded out in the 1980s. It seems to have been replaced with more ambiguous concepts of 'soft science'. The USA institutionalized technology assessment in the US Congress by establishing the Office of Technology Assessment, through which the concept has been perpetuated. For the Japanese, it is not an endogenous concept but merely an imported catch word destined to be replaced by a new import.

Establishment of Environment Agency

The establishment of the Environment Agency in 1971 was nothing but an outcome of a response by the public sector to the rising pressure of the citizens movement, though the public sector may be reluctant to admit it openly. In the same year, many prefectures started prefectural research institutes to study pollution problems. In that year, the public R&D budget for pollution measures continually rose[18] until the time of the oil crisis in the fall of 1973. The Japanese bureaucracy had so far been slow in increasing public expenditure on welfare and had not been exposed very much to citizen pressure. But the creation of the Environment Agency was an exceptionally quick response to popular demands. The American Environmental Protection Agency, which was established in the preceding year of 1970, established a model which could be emulated. The Japanese Agency was so named because of its aims of not only preventing pollution but also to protect the natural environment.[19]

Each government agency corresponds to some profit-oriented enterprise or interested groups. Each branch of MITI represents the interest of each industry concerned. In this respect, the Environment Agency is really an exceptional agency, which has no support from economically interested groups but only the people at large, whose opinion is represented by the mass media. At least in the early stage, the Environment Agency advocated humanitarian values, that could not be reducible into economic terms.

The ideology of the Environment Agency, the staff of which are often seconded from the Ministry of Health and Welfare[20], was diametrically opposed to MITI (and STA) which claims that even environmental problems should be brought within the scope of economic rationality, that is to say, everything in its final analysis can be reducible into a costbenefit analysis. In MITI's way of thinking, it is the technology that created environmental problems and it is only the technology that can solve this problem. This is often criticized as being a technological fix or 'match and pump' logic. MITI seeks solutions for problems through the collective economic efforts of all the enterprises, namely by creating new enterprises to specialize in solving pollution problems for which other enterprises are responsible. Some new enterprises have even claimed to have taken advantage of Japan's congested environment to work on pollution-prevention measures, with the hope of exporting this new technology in the future to less congested and hence less advanced areas such as the USA or Australia. On the other hand, the Environment Agency from the very beginning criticized MITI's logic and requested a change in their way of thinking.

This controversy was not long-lasting. The support of the mass media for anti-pollution movements was short-lived, as its news value faded. During the 1970s, when the power of the movement began to decline, the Environment Agency had to retreat under pressure from MITI. Those radical bureaucrats, who helped to form the opinion of the Agency in the beginning were replaced by less radical staff, step by step[21], and finally there have been occasions where the Agency has supported the private sector against citizen opposition.

Response of Public Sector Laboratories

The role of public laboratories has been a matter of controversy in science policy. In the late nineteenth century when the Japanese government embarked on Westernization, there were a number of state service-science endeavors such as the establishment of meteorological and astronomical observatories and geological and land-survey networks. Since the early twentieth century, the Japanese turned to creating public-sector industrial laboratories, which helped Japanese industries in providing scientific and technological information and R&D. That is, they helped to provide services to industrial sectors before the war, when Japanese industries were still unable to have their own laboratories.

In the postwar period, MITI's inhouse laboratories were expected to provide basic science to industries, and liaison between academic and private sciences. In the early 1970s, the anti-pollution movements 'brought pressure to bear upon them'. Those top research managers, who had so far set the goal of research teams for the service of the industrial world, now lost self-confidence. Then, rank and file researchers in the laboratories had to rethink their goal and role by themselves, freed from top-down target setting; some of them looked for alternative science or appropriate technology.

Private laboratories can do profit-oriented private science better than the public sector laboratories and the former came to be powerful enough to conduct the basic science which they required by themselves. The target of public sector laboratories, should then lie in service science for citizens, such as microelectronic devices for disabled people, as practised in a MITI laboratory. Such research is not sufficiently profitable for private laboratories.

Academic Response

A significant contribution of the academic sector has been the establishment of the discipline of environmental science. Born out of the needs of the time, environmental science enjoyed a period of Ministry of Education support around 1973 which saw the establishment of new courses, new departments and even new schools in various universities.[22] Nevertheless, there were not many applications for research funds on the subject. This meant that the new environmental science was not able to generate an independent paradigm. Textbooks were created which merely juxtaposed the practical arts of prevention of water pollution, air pollution, hazardous pesticides and so on with an introductory ornament of the science of ecology. Thus it failed to develop into an academically respectable subject. Hence during the 1980s, many environmental science courses in universities were disbanded or renamed. Only in the late 1980s, when issues on environmental crisis have appeared on a global level, have scientists started to reconsider the matter again seriously.

Response of Private Sector to Pollution

During the previous high economic growth period, the private sector often led national opinion. For instance, the Keidanren (Federation of Economic Organizations, the representative body of the Japanese industrial world) advised the government to increase scientific manpower in August 1961. The government immediately responded and its Ministry of Education formulated and announced in the following month of September a plan of increasing 20,000 students of science and technology by 1964. Consequently, schools of science and technology in government universities expanded to twice their former size and many private universities opened schools of science and technology, which were hitherto considered to be difficult to support in the private sector. And finally, the so-called science and technology boom started nationwide, thus completing a topdown flow of opinion, from private, to public, to academic and to citizen sectors. This process was reversed in the phase of reflection and rethinking of science and technology in the late 1960s and early 1970s.

Its earliest manifestation may be found in September 1964, when residents' opposition was organized against the construction of a petrochemical plant in the city of Numazu, and the city council voted against it. University scientists identified the relationship of industrial waste of methyl mercury with Minamata disease in 1963 and *itaitai* disease in 1965. The word *kogai* (pollution, literally public nuisance) appeared in the newspapers everyday. In 1965, the Pollution Council (Kogai shingikai) was formed as a government advisory committee. In August 4, 1966, the Council sent an interim report to the Minister of Welfare to emphasize the responsibility of the government and enterprises. The Council drafted the Countermeasures Basic Law (Kogai Taisaku Kihon Ho), about which the Keidanren protested as being too harsh for enterprises. Finally on August 3 in 1967, the Diet enacted the law, which set standards of pollution. MITI insisted on modifying the law drafted by the Ministry of Welfare. The

Keidanren complained about the validity of the law from time to time but gradually accepted the antipollution measures. MITI and Keidanren continued to experience 'hard times' in terms of pressure to adapt to changing attitudes towards the environment.

The oil shock in late 1974, in a sense, saved them and revived their morale. The energy crisis and ensuing inflation turned mass media's attention away from *kogai*. Taking advantage of this new trend, the private sector tried to recover some of the influence and power which it had enjoyed before. The industrial sector gradually changed their industrial structure from heavychemical to lightmicroelectronic industry, or as the popular slogan goes from 'heavy thick long large to light thin short small', so as to minimize both pollution and energy use.

Alternativism

While major events relating to science, technology and society have been student protests and university reforms in 1968–70 and environment and ecology in 1972–73, book publication statistics show[23] that the attention of the citizen sector has been directed towards resources and energy and alternative lifestyle since 1974, as a result of the oil crisis which occurred in the fall of 1973. Publications on nuclear energy briskly increased in 1975. Since the themes imply a different worldview (or cosmology in ecologist terminology) and discussion of lifestyle issues, the problems have not been easily resolved to date. In the following, we shall illustrate some of the points of debate.

technology	conventional	alternative
management	centralization	decentralization
scale	capital-intensive	labor-intensive
energy	fossil	recycable
workers	specialist	do-it-yourself

While in the beginning of the critical reflection period in the 1970s, the energy of the citizens sector was spent criticizing the behaviour of government and industrial sectors, by the mid 1970s a movement had emerged with alternative proposals to build up their own type of science and technology, which questioned and confronted the existing routine of technocratic science. Alternativism excercised considerable influence in various fields like education, medicine, economy and politics, from which the alternative technology (abbreviated as AT) movement has gone on to survive until this very day. In short, it is nothing but 'another technology' more decentralized and manageable by the citizenry by themselves, challenging the big, centralized, high technology developments controlled by the established mainstream technocrats.

The AT movement continued to spread, though, losing its political radicalism which had existed at the outbreak of the movement. This could be interpreted as a shift from left wing focus on political revolution to a concern for a revolution in lifestyle as represented by the ideas of the Green Party of West Germany. Because this was not like a violent revolution, it did not attract much attention. But it has exerted invisible influence on the lifestyle and sensitivity of the citizenry through such tangible and concrete ways as the popularization of organic farming and the do-it-yourself movement.

Japanese alternative movements in the citizen sector may be, however, less visible than their counterparts in the Western countries in Europe, USA and Oceania, as they are overshadowed by the more impressive Japanese technocratic system. This system was endangered at the time of the 1973 oil crisis that hit the Japanese economy more so than anywhere else in the world for the technocrats successfully turned the attention of people from pollution to energy needs. Government and the private sector took advantage of the crisis and minimized their efforts for ongoing pollution prevention. While R&D expenditure for environment protection remains nearly unchanged since the mid1970s, those for nuclear energy and information technology have enjoyed rapid increases in funding.[24]

Alternative Energy Strategy

On the other hand, crudely similar to the citizens alternativism has been the public and private sectors' response to the oil crisis and their attempts to create an alternative energy source project. In the post-oil shock period there began a search for natural resources of energy like solar and windpower etc. MITI's 'Sunshine Project' (solar energy project since 1974), and 'Moonlight Project' (energysaving project since 1977), and the Ministry of Agriculture and Forestry's 'Green Energy Project' (since 1978) are typical of them. Indeed, during the 1970s, Japan scored the world's highest percentage of government R & D funding for energy problems.[25]

This kind of energy plan naturally changes according to the demand and supply situation and the character of the political system. In the US, during the Carter administration there was a tendency toward solar energy, but during the Reagan administration the weight has shifted toward nuclear technology. The change in research and development funding was not limited to new energy projects. In US, Britain and France, large scale budget

Figure 12 Portion of R & D in government expenditure

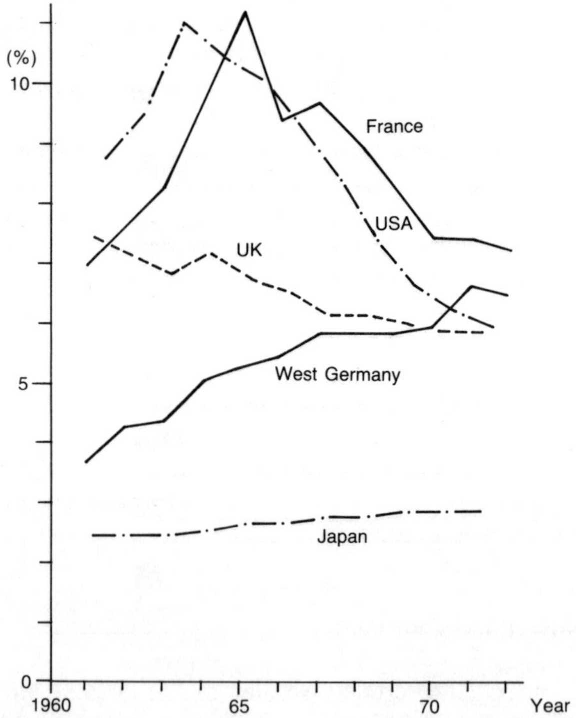

Source: OECD Committee for Scientific and Technical Policy, *Trends in R & D in the Higher Education Sector in OECD Member Countries since 1985 and their Impact on National Basic Research Efforts* (Paris: October 15, 1979)

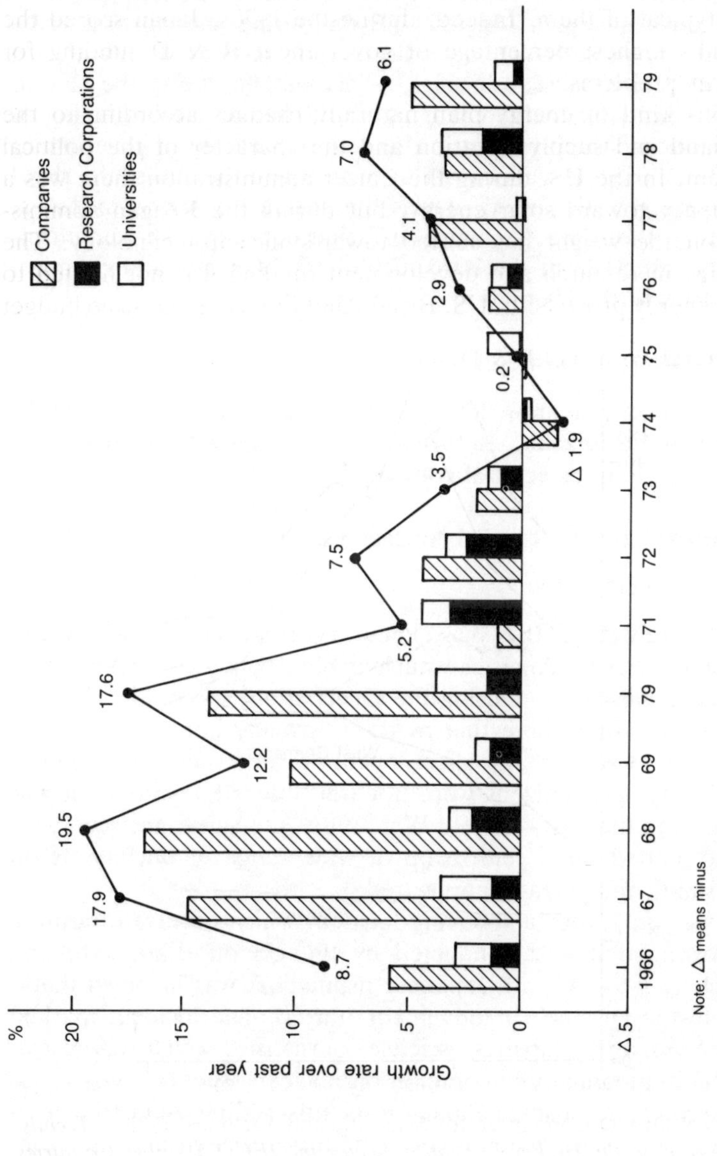

Figure 13 Growth rate of Japanese R & D expenditure
Source: STA, *Kagaku Gijitsu Hakusho* (Science and Technology White Paper, 1981)

cuts in R&D have occurred since the late sixties. But funding in West Germany and Japan, though much lower than major Western countries at the beginning of the postwar period, has seen continuing increases (fig. 12). Even during the post-oil crisis period, Japanese R&D expenditure soon recovered only after a year of decline as seen in fig 13. This may be due to the lack of policy sensitivity to the changing world due to the hardening of the financial structure, but it also reflects stability in government R&D planning. If we look at its content with the end of the rapid economic growth period, it was the search for a new industrial structure that forced change in science policy. The emphasis clearly shifted from the hard technology of heavy and chemical industries to the promotion of life science.

SECTION 2. Against Nukes

Nuclear energy is now at the focal point between the establishment and the citizenry. In this section, we deal with the scientists' stance in the intersectorial controversy.

Scientists' Attitude Toward Nuclear Energy

Toward Nuclear War

Immediately after the War, even the renowned leftist, Mituo Taketani, viewed American nuclear physicists as liberators from Japanese militarism. During the Occupation, the press code had created an atmosphere that made it virtually impossible for the Japanese to raise their voices openly against nuclear weapons. Hence Japanese citizens were not well informed as to what was going on in the postwar Cold War. When the San Francisco Peace Treaty ended press censorship, a wide range of documents on Hiroshima and Nagasaki appeared.[26]

It was only after a shocking incident which occurred during a hydrogen bomb test conducted by the US on Bikini Atoll on March 1, 1954, when a Japanese fishingboat was involved that a nationwide antinuclear movement started. The incident marked the beginning of Japanese science journalism, which was instrumental in guiding public opinion on nuclear issues.

There are no postwar Japanese scientist militaryindustrial complexes established which involve military research and therefore there are none to defend their position against popular antiwar attacks. Whether explicitly or not, all Japanese scientists are against war and particularly against nuclear bombs. There is, however, a prolonged controversy between the JCP and other parties, notably the Socialist Party, on the kind of bombs to oppose. The controversy may go back to prewar peace movements, such as in the USA, in which communists legitimized the war against capitalist countries, while socialists took an absolute pacifist stance against any kind of war. In the postwar period the JCP claimed the righteousness of the Soviet nuclear bomb, while the Socialist Party opposed those of all countries regardless of ideological stance.

Communist-affiliated scholars missed the point, however, that the Soviet bombs were also promoted by a technocratic structure which was rooted in a Soviet militaryindustrial complex, such as existed in capitalistic countries. Soviet and American technocracy has been in competition with each other but without the existence of an opponent these technocratic structures cannot maintain their own raison d'etre.

A. Scientists at War – Kyoto Scientists Conference

Early in the postwar period, elite scientists were able to lead public opinion. One symbolic example is found in the Kyoto Scientists Conference (Kyoto kagakusha kaigi), the Japanese version of the Pugwash Conference headed by Japan's first Nobel laureate, Hideki Yukawa, and supported by his pupils and associates. It was generally felt in the postwar Western world that scientific elites, especially physicists, could foresee the future world better than political professionals, since elite physicists were not yet incorporated into the technocracy and hence behaved as if they were free independent thinkers. This is the spirit which supported the ideal of the Pugwash Conference as well as the Kyoto Scientists Conference.

Yukawa was one of eleven original cosigners of the Russell-Einstein manifesto urging the abolition of the bomb in 1955 and also one of seven original promoters of the Pugwash Conference, which started in 1957. Yukawa was frustrated with the subsequent trend of the Pugwash Conference stance as it moved away from its original spirit of nuclear disarmament and approached a stance of strategic nuclear deterrence. Yukawa wished to restore the original intention of the Pugwash Conference by inviting members to Kyoto with his own initiatives; this was named the Kyoto Scientists Conference. It took place for the first time in May 1962 and then in 1963, 1966 and 1981, the last year of Yukawa's life.

Both the Conferences have no intention of appealing to the citizen level but rather to internationally renowned elites of Nobel laureate standing, with the exception of the Marxist Shoichi Sakata who tried to appeal to citizen organizations. The Kyoto Scientists Conference was discontinued after the death of Yukawa, while the initiative of peace movements has moved from elites to citizens.[27] The death of Yukawa also signifies the demise of the nontechnocratic scientific elites.

Atoms for Peaceful Use

Japanese scientists are in a pacifist position in dealing with nuclear science and technology. The postwar constitution prohibited Japanese participation in the war. In relation to nuclear research, some Japanese scientists interpreted this constitution to mean that Japan was able to work only on the peaceful application of nuclear energy, namely nuclear reactors, while staying out of involvement in research for atomic bombs. A difference of opinion and strategy has emerged as to what constitutes 'peaceful use' in a later controversy as explained below. The case is interesting as it shows the standpoint of the scientific community in a dichotomy of democracy and technocracy.

1. Towards Nuclear Reactors – Three Principles of Nuclear Research

In the United States, research into atomic energy started with the bomb and branched off to reactors for peaceful application. Both military and peaceful applications share a common infrastructure, research personnel, technical knowhow, and classified information. Hence, even for peaceful R & D the principles of academic science may not necessarily take priority. The principles of industrial and commercial science do not always dominate either, for the products of nuclear research are not assessed by the market. It is a pseudocivilian science, which is conducted in secluded circumstances with little assessment by the people.

The case of Japanese research is quite different from the above. In April 1954, at the outset of Japanese research, the Japan Science Council proclaimed that in order to keep nuclear research along peaceful lines, atomic research should be conducted along the three principles of 'autonomy', 'public accessibility' and 'democracy'. These principles originate from academic science, and diametrically opposed incorporated science, as practised in nuclear research elsewhere.

There was considerable apprehension amongst Japanese scientists, particularly amongst the Minka faction of the JSC that once atomic energy research was commenced, there would be a tendency for research to be classified. In order to alleviate their anxiety, the three principles were laid down and the Japanese government adopted them into their Genshiryoku kihonho (Basic law of atomic power) which was created the following year. They were optimistic enough that the academically established rule would be followed without question by other sectors.

Internationally, however, Japanese scientists who made study trips abroad realized that in the 'real world', these three principles were out of the question. 'Public accessibility' simply had never existed in the short tradition of nuclear research, whilst the principle of 'autonomy' was infringed by the pressure of export marketing by the American and British nuclear establishments.

At this time, Japanese physicists, who belonged to the Science Council of Japan (JSC), led by Mituo Taketani, took precautions against the misuse of imported reactors on the basis of the autonomy principle. The Japanese scientists, who were now permitted to work in the once forbidden area of atomic power, had insufficient experience in constructing reactors but they turned out to be severe assessors of foreign made ones. Taketani commented that 'since atomic power will be a determinant for our future life, we should not be rash and thoughtless in dealing with it. We should closely adhere to these three principles and patiently accumulate research data under the guidance of physicists rather than technologists, for to do otherwise will invite grave consequences.[28]

In 1957, an accident occurred at a British Windscale reactor for the first time in the history of nuclear reactors. At that time, the Japanese government agency was scheduled to import the same style reactor and so a controversy arose as to whether to import the reactor or not. While physicists of the JSC were highly critical, political and financial establishments were only too happy to start new enterprises around the imported reactors. Technologists on the side of the Genshiryoku iinkai (Atomic Energy Commission) and the STA as well as the industrial world were concerned about falling behind international trends in atomic energy.

At the time, a controversy arose in the scientific community of whether or not Japan should import nuclear technology or to develop its own and remain autonomous. While the academic sector, notably the physics community led by leftist ideology insisted on autonomy, the emerging technocrats chose the pragmatic line of importation, upon which future autonomy could be established.

2. Three Standpoints of Scientists on the Matter of Reactors

There are three schools of opinion regarding the promotion of nuclear energy: 1. energy, 2. science, 3. ecology. Each of these 'schools' roughly corresponds to the following groups: 1. technocrats, 2. unionists, and 3. populists respectively, while ordinary academics remain silent.

a. Energy School

The views of this school are based on the premise that in order to maintain the pace of present growth it is imperative to secure sufficient energy sources. In view of the limited petroleum resources, it is a logical conclusion, they feel, to switch to nuclear energy. They are the people who are responsible for conserving the economic and political status quo for the sake of the existing establishment. They, as occasionally found in the post-oil-crisis argument in the mid 1970s, tactfully confuse the energy source for tomorrow with the longterm trend of available world resources and promote the necessity of nuclear reactors as the only way for us to survive on Earth.

b. Science School

Their view is that although atomic energy research is in the long run absolutely necessary, reactors have yet to be developed to the stage where practical use can be made of them. Therefore, they say, we must spend more time and funds on research. As scientists, they criticize the standard of current safety measures, disclose the technical deficiencies of reactors, and warn against possible hazards. They suggest a moratorium on nuclear reactors as a tentative possible solution. This is the standpoint of scientists critical of reactors; this attitude is compatible with and shared by the main stream of JSC and NKK (Japan Scientists Association), who believe in the eventual usefulness and possibility of peaceful nuclear energy.[29]

Their critical activity was particularly visible and significant in the late 1970s, when Japanese reactors had the most frequent troubles and accidents.[30]

c. Ecology School

This school considers that nuclear hazards may be prevented in the short term by surveillance by 'science school' scientists, but because of ecological considerations, reactors cannot be the ultimate solution for future energy problems due to nuclear waste. They conclude, therefore, that reactors should be abandoned. This is the standpoint of an ecologist who is leading an antinuclear campaign of community residents.

The ecology school suggests that it is unnecessary to unleash excessive energy, for to do so would invite military uses and the destruction of nature. Hence, they argue, we should not harness more energy without having a clear and legitimate plan for its use. This is the claim of an ecology fundamentalist.

Since 1962, the Japanese economic establishment has switched its energy source from traditional coal to cheap petroleum imported from the Near East in great quantity; the need for atomic energy was temporarily postponed. Nevertheless, a controversy between nuclear technologists (the energy school defined in the following section) and physicists (science school) continued and peaked in the intense debate of the 70s.

After the oil shock in 1973, the new demands for energy accelerated the planning and construction of reactors, even though it did not meet the demands for energy created by the immediate shortage of oil, as it takes five to ten years from planning to operation. At the same time, however, there arose a number of unprecedented and inexplicable accidents, which frustrated the promoters of nuclear energy. These events added fuel to the fire of ever growing opposition, which continues into the 80s.

3. Debate Turns from Technical to Ideological

On the matter of the reactor debate, the science school has tried to maintain a neutral and critical attitude to the Three Mile Island and Chernobyl incidents. It is, however, more than a piecemeal technical debate. Major Japanese debates are on the issue of whether reactors as planned by technocrats in Tokyo should be built at a remote area by evacuating local residents. Problems are social and political rather than technical in nature. It is a matter of total ideological and structural differences between technocracy and democracy, in which scientific expertise has not much to say. Some members of the science school still maintain that the peaceful application of nuclear energy should be promoted but that too hurried an application should be avoided until adequate research is completed. They have repeated this claim for more than four decades. Taketani, once the leader of this science school, now suggests that the transition of the Japanese industrial structure from heavy to light industry means that there is no real need for nuclear energy. The science school will eventually be split and absorbed by either the energy or ecology schools.

Citizen Opposition to Nuclear Reactors

During the 1950s, there was no notable opposition to reactors. In the 60s, opposition was only locally organized at the site of reactors, usually at remote locations along the coast, where a great deal of money was spent on compensation for the loss of the livelihood of fishermen. Only in the 70s under the critical science movements, was nationwide opposition organized and the Socialist Party, the largest opposition party and the party that represents the interests of trade unions, has committed itself to a moratorium on reactors, though their policy has occasionally wavered as they have met with disagreement from the utility and electric manufacturer's unions.

After the Chernobyl incident, opposition has rapidly spread amongst housewives who have grave concerns about food pollution. The radiation fallout from the Chernobyl incident in April 1986 was detected even in Japan. Since then, like everywhere else, the Japanese citizen movement against nuclear reactors has gained strength, particularly involving concerned housewives. Though the movement itself started much earlier, the Japanese nuclear establishment for the first time took the opposition movement seriously and spent a considerable amount of funds for public relations using the mass media. Although they do not forget to emphasize the incomparable safety measures of Japanese nuclear technology from the science school viewpoint, they find difficulty in persuading people of the need for nuclear energy on the grounds of energy school logic, as their energysaving projects since the late 1970s have proved to be unexpected successes and oil consumption actually declined. Instead, they must find grounds for argument in a global increase in energy usage, particularly in the Third World.[31] The citizen movement has been led and advised by scientists of the ecology school in opposition to scientists in the nuclear energy establishment. Public hearings have been often conducted but confrontation continue without any sign of compromise.[32]

Notes

1 Nihon kagakushi gakkai (History of Science Society, Japan, ed.) *Nihon kagaku gijutsu shi taikei, tsushi* (Compendium on the history of science and technology in Japan, Outline History, no.5, 1964, Daiichi Hoki) propounded this approach.

2 From Minshushugi kyoiku kyokai (ed.) *Daigaku gakusei mondai bunken mokuroku* (A catalogue of literature on university and student problems, 1965–1971 (1971)).

4 *Ibid.*

5 Parrallel to the internalexternal dichotomy as employed by the historians of science, I use hereafter a dichotomy of the externalist critique on the socioinstitutional structure of science and the internalist on the ideological side.

6 Shigeru Nakayama, *Kagaku to shakai no gendaishi* pp.110–113.

7 Tetu Hirosige, Atuhiro Shibatani, and Jinzaburo Takagi are the early proponents of this new perception of science. Hitoshi Yoshioka and I have recently edited a journal focussing on this. Shigeru Nakayama, and Hitoshi Yoshioka (eds.), 'Taiseika kagaku no kozo bunseki' ('Structural analysis of technocratic science'), *Rekishi to shakai* (*History and Society*), vol.9 (April, 1989, Libroport).

8 Tadao Uchida and Shinkichi Eto (eds.), *Atarashii daigakuzo o motomete* (Call for a new image of universities, Nihon Hyoron, 1969).

9 Kazuyuki Kitamura's estimate in Shigejiro Tabata *et. al.* (eds.), *Nihon to gaikoku no shodaigaku kaikakuan* (*University reform plans in Japan and abroad*), (Yushindo, 1971).

10 Nakayama, *Kagaku to shakai no gendaishi*, pp.93–99.

11 Shigeru Nakayama, *Kagaku to shakai no gendaishi*, p.99.

12 Nakayama, *Kagaku to shakai no gendaishi*, pp.12–13.

13 Jun Ui, 'A basic theory of Kogai', in S. Nakayama et. al. (eds.), *Science and Society in Modern Japan* (University of Tokyo Press, 1974) pp.290–311.

14 Hoshino, 'Sengo gijutsushi', pp.101–102.

15 The Soshisha (Institute of Compassion) at Minamata, which was a onetime centre of antipollution movements is even now manned by the generation of students who migrated from universities to Minamata in the 1970s.

16 Robert Jungk, 'Technological forecasting as a tool of social strategy', in R. V. Arnfield (ed.), *Technological Forecasting*, (Edinburgh U.P.).

17 *Gijutsu yosoku*, (Hara Shoten, 1968).

18 *Kankyo hakusho* (White paper on environment, ed. by Environmental Agency, 1972) p.61.

19 *Kankyocho junenshi* (Ten year history of the Environment Agency, 1981), p.64.

20 *MITI*, pp.7778.

21 Hideo Otake (ed.), *Nihon seiji no shoten* (Focal points in Japanese politics, 1984), pp.233ff.

22 Akira Tezuka, 'Kankyo kagaku kenkyu no naritachi to kongo no tenbo' ('The establishment of environmental science research and its prospect'), *Gakujutsu geppo*, vol.30, no.7 (1977), pp.478–485.

24 *Kagaku gijutsu hakusho* (Science and Technology White Paper, 1988) p. 129.

25 *OECD Science and Technology Indicators Resources devoted to R & D* (OECD, 1984) p.87, Table 2.9.

26 Toshiyuki Toyoda, 'Japan's policies since 1945', in L. Ackland and S. McGuire (eds.), *Assessing the Nuclear Age* (1986, Educational Foundation for Nuclear Science, Chicago) p.58.

27 Shigeru Nakayama, 'Kyoto Kagakusha Kaigi', *Rekishi to shakai*, no.6 (1985), pp.144–168.
28 Taketani's phrase recapitulated by the author.
29 Shoichi Sakata, *Kagaku to heiwa no sozo* (Creation of science and peace, Iwanami, 1963).
30 *Genshiryoku anzen hakusho* (*White paper on nuclear safety*, ed. by Japanese Commission on Nuclear Safety, 1988) p.3.
31 Genshiryoku iinkai (Japanese Atomic Energy Commission) ed., *Genshiryoku hakusho* (White Paper on Atomic Energy, 1988), p.56.
32 Genshiryoku iinkai (Japanese Atomic Energy Commission), ed., *Genshiryoku hakusho* (White Paper on Atomic Energy, 1988) pp.76–78.

8 Microelectronics Revolution

Japanese technocrats are very proud of their role in the development of microelectronics. Work in the area was initiated in the public sector and later handed down to the private sector for commercial development. The social cost of such technology was recognized in the public sector only after international concern about its impact on society had become apparent. In this chapter, we focus on the interplay of three social sectors, public, private and citizen. It is significant to note that the academic sector was almost entirely removed from the scene in the 1980s.

SECTION I. The Impact of New Technology

In the present framework of society, new science and technology is promoted by a technocracy and assessed by a democracy. New technology is the source of major social change in contemporary society. As to whether such change is judged good or bad depends upon the value systems within the different social sectors. The new technology is developed in the technocratic, public and private sectors for the benefit of each respective sector. When the influence of the technology reaches other sectors, assessment will almost always vary between sectors, though the long term consequences of the new technology are rarely immediately foreseeable. In such a case, the critical assessment by the citizenry and academic sectors are legitimated, their viewpoints often being represented by science journalists and critical scholars. The academic sector will usually attempt a neutral stance in the debates, but close scrutiny inevitably discloses political leanings.

The social function of a critical, and sometimes pessimistic, academic assessment is a warning of possible consequences rather than an outright prediction of outcome. Consequently, such an assessment should be evaluated not in terms of accuracy of prediction, but rather through its power as an important political tool to promote the needs and desires of the citizenry sector over the economic rationalists in the technocratic camp. Recent debates on microelectronics are a typical example.

The Microelectronics Revolution

Around 1978, the advent of the new technology of microelectronics caused a flurry of debate among science policy makers in Europe concerning this new phenomenon, the 'Microelectronic Revolution'. The present conceptualization of 'new technology' usually includes biotechnology. However, in the 1970's, biotechnology was only beginning to have an impact on the world scene.

The term seems to have originated in the UK. The British and EC establishments intended to impress their people by calling microelectronics R&D a revolution, a revolution which had been predicted to be the source of the largest technological innovation in the next quarter century. Such aggrandisement suggests a crisis consciousness within these countries of being behind the USA and Japan, and a fear of being excluded from a share of the world market. In the UK, the slogan 'microelectronics or bankruptcy' appeared.

Microelectronics is, of course, not a new technology that appeared in the late 1970s. It goes back to the prewar research of William B. Shockley, from which came the transistor which was officially patented in 1948. Since then, the US government has invested a great deal of public money on computer R&D for military and space purposes and in the process has established itself as the outright world-leader in IC (integrated circuit) and LSI (large-scale integrated circuit) development. In the third quarter of this century, almost all important inventions and innovations in semiconductor devices have been made in the United States.[1]

While dependent on American frontier research, Japan in the meantime sought to compete with the USA in the development of commercial and consumer goods such as radios, TV sets and tape-recorders. Japan started to produce ICs in the mid-sixties and overtook the USA in the total sale of consumer electronics in the mid-seventies[2] In the same period, it became quite obvious that Britain and the EC were far behind in the development of microelectronics, as they adhered to vacuum tube technology even when transistors became available, only starting to produce ICs around 1970.

MITI's strategy

From the 1970s, MITI recognised that Japanese technology of the type characteristic of semi-developed countries, such as shipbuilding and textiles, came to be matched, sooner or later, by NIEs such as Korea and Hong Kong. This being the case, MITI decided that Japanese technology would only develop when it outstripped the high technology of the USA. Consequently, MITI decided to encourage changes in the Japanese industrial structure to align it with that of the United States. The most notable American technology at the time was the high technology of aeronautics and microelectronics. Japan was not able to compete with American aeronautics, since the U.S. aeronautics industry has been afforded military support, a situation which is constitutionally difficult in Japan. Hence, MITI decided to concentrate on microelectronics. To achieve this aim, MITI provided low-interest loans from 1971 to 1978 under the Electronics Industry Promotion Temporary Measures Law, in an effort to encourage the development and manufacture of products such as numerically controlled machinery and industrial robots.[3]

MITI also generously subsidised microelectronics R&D,[4] with such schemes as the VLSI development project, started in 1976, with a considerable slice of the national budget. Some 74 billion yen was spent, about 40 percent of MITI's R&D funds. This was a most unusual move since, in most cases, it was postwar Japanese practice for R&D funding to hail from the private sector. As a result of this national initiative, the project came to be nicknamed the Rising Sun (Hi-no-maru) computer project.

MITI's typical method of support is as follows: MITI invites specialists from several different corporations, including the academic sector, and proceeds to negotiate with them on organizational arrangements. MITI then creates a technology research association[5] and subsidizes it by establishing the funding device of an auxiliary organization to enable flexible expenditure of public money[6]. Once R&D reaches the point at which business can be competitive in the marketplace, usually after a period of a few years, the association is disbanded and MITI leaves further development to market competition. In this way, Japanese levels of technological development came to approach American standards.

Thus it came about that the USA and Japan both commenced competing national R&D projects for VLSIs. Four years after its inception, the Japanese VLSI research association was officially terminated in 1980, the year in which the Japanese outstripped the US in semiconductor technology. Due to the success of the Japanese VLSI project, G. Gregory claims that popular perceptions abroad, which cast Japanese industry in the role of imitators, clearly were outdated.[7]

Several years behind Japan and the USA, European governments came to recognize the wisdom of such an R&D policy and tried to get into the game in the late 70s. The UK and France initiated national VLSI R&D in 1978. The UK and West Germany began to spend public funds on microelectronics comparable to the expenditures of the USA and Japan from 1979.

Negative aspects of the Microelectronic Revolution

The British Department of Industry commissioned a technological forecast report, published in January 1978, in an effort to encourage promotion of the microelectronic industry. Its conclusion stated that it is more important to formulate socio-economic measures to prevent things such as unemployment rather than a technology policy to promote a computer industry. The Department of Industry refused to publish this report, but since then the content of the report has provided the basis of the subsequent debate on the social impact of electronics.[8]

There are three conceivable major types of negative social impact of computers:

1. unemployment
2. work alienation
3. infringement of privacy in a controlled society.

1. The threat of unemployment was, without delay, taken up by labor unions as the issue of greatest concern. Demands for the

shortening of work hours and increase of paid holidays were a regular feature in line with their ongoing anti-rationalization campaign. 2. Work alienation problems are also a matter of negotiation between management and unions but it is rather difficult to manage in terms of traditional claims by labour unions; it is a new problem to be dealt with by new-left oriented rank and file workers with the support of ecologist groups and the Green Party. 3. The problem of privacy is argued in the wider social context. It is largely a matter of choice between a centralized or decentralized social regime.

The Advent of the Leisure Society?

Microprocessor's Influence on the Tertiary Industrial Sector in the West

'Micro'-electronic components are cheap and microscopic. Their effects tend not to be quite as obvious as those of the products of macroengineering. As a result of their microscopic nature, however, they are able to penetrate every corner on earth and pervade every aspect of society.

Perhaps the most conspicuous side-effect of the advent of microelectronics is unemployment. The introduction of industrial robots containing microcomputers into the manufacturing sector accelerates production but also invites mass-unemployment in secondary industry. The permeation of microelectronics into tertiary industry via office automation, threatening redundancy of typists, secretaries, clerks and even managers is a less noticeable development, but perhaps more important in terms of the potential levels of unemployment within this industry. Office automation made its debut in Western countries where typewriters came to be widely used. Through the simple swapping of typewriters for word-processors, secretarial work can be cut by half. From this, it has been reasoned that the office workforce will be halved.[9]

Perhaps the most radical influence of microelectronics took place within the printing industry. With the introduction of word-processors, typesetters became largely redundant. As a result, the anti-rationalization struggle was frequently in effect in the early 1980s, such as the long term press strikes experienced by the *London Times*. To what extent unemployment is attributable to the introduction of microelectronics and to what extent to other conventional causes is impossible to determine.[10] In reality, the effect of microelectronic unemployment has arrived slowly and gradually compared to that due to rapid economic depression and recession. Technological unemployment, however, appears to be expanding in a oneway direction, as distinct from the cyclic nature of the unemployment caused by economic depression. In spite of economic growth in terms of GNP, the unemployment rate is ever-increasing. Such economic prosperity has been called 'growth without work.'

A number of studies which predict the possible effects of microelectronic unemployment have appeared since the late '70s. For instance, a French semi-official report, the Nora and Minc report[11], has said that within ten years, 30% of the labor forces of the present secondary and tertiary industries will be redundant. A more extreme prediction is that of the British economist T. Stonier, who predicted in 1979 that by early next century we will require only ten per cent of today's labour force to provide all our material needs.[12]. Despite variation in the numerical values of predicted obsolescence, all of the critical appraisals predicted the certain decline of employment with the usage of micro-electronics as labour saving devices, as a course of logical necessity.

In an effort to placate the labour sector, those technocratic advocates for the introduction of microelectronics into the workplace, were assured that new occupations, such as jobs in software, could be created to compensate for microelectronic unemployment. As long as economic growth continues, such a solution of unemployment may be possible for some time, but in the long-term, ever-growing unemployment would seem to be a logical consequence unless serious countermeasures are instituted.[13]

The arguments described above are mainly taking place in Europe. Since 1978, numerous books and articles have been published on the nature of the microelectronic revolution and its possible future prospects. Opinion is divided into two camps-optimism from the technocratic establishment and pessimism from critical science policy makers (whose views have even been tabled in reports to the government) and social scientists. Exemplary work may be found in the report of the Committee of Science and Technology Policy of the OECD entitled *Science and technology in the New Economic Context.* British economists such as Christopher Freeman of the Sussex Science Policy Unit and D. Leach and H. Wagstaff, are interested particularly in the virtually irreducible percentage of Britain's unemployed.[14]

In contrast to European debates, American and Japanese reaction is slow and still very much in the background. The contrast in levels of interest can be explained as follows: while the USA and Japan are producers, exporters and invaders of the world markets of microelectronics and thus its beneficiaries, European countries, the latecomers in the trade, remain solely importers and victims of the microelectronic revolution. It can be explained succinctly with the Japanese proverb, 'a rich man never quarrels.'

Effect of Word Processors in Japan

When European critics initiated debate on the negative aspects of the microelectronic revolution in the late 1970s, the Japanese general public was not aware of the issues being discussed. One of the many reasons for this obliviousness was that, because of the ideographic nature of the Japanese language, typewriters were not widely utilized in Japanese offices and the average person did not believe in the prospect of Japanese word-processors. With the advent of VLSI, the situation changed[15].

In September 1978, Toshiba announced the first Japanese word-processor with a kana-kanji transformation device and in 1980, planned to sell 1,000 word processors in the domestic market; in the following year of 1981, a rash of new products were introduced by other Japanese companies, Canon, Hitachi, Matsushita, NEC, Nippon Univac, Sharp and Ricoh.[16] Since then, Japanese word-processor sales have grown at an average rate of 47.7 percent over the 1983–88 period.[17]

And so it was that, a decade later than in the West, word-processors became widely available in Japan through a successful conversion from phonetics to ideographs. Such technology was considered revolutionary since it was more than the conversion of the typewriter to microelectronic writing; it was the transformation of handwriting to microelectronics. Japanese wordprocessors with the capacity to type thousands of different characters would be a very elementary AI (Artificial Intelligence) or Expert System, if it is defined as 'a computer program to solve problems that would otherwise require a human expert.'[18] It fits the claim made by the advocates of the 'expert system', which is defined in the following way.

> In general there is a shortage of people in industrialised society who have advanced knowledge of things, while there is an abundance of those who do not. Therefore the value of portable knowledge in the shape of an expert system is that of making skills available to a wider population. This is surely an up-skilling property and not one that robs humans of their toil.[19]

Before the introduction of word-processors, it was standard practice in Japanese offices to communicate with handwritten letters. Only exceptionally important documents were typeset by specially trained expert typists with Japanese typing machines that cannot be handled by ordinary clerks.

Since the introduction of microprocessors in Japan, Japanese typing machine typists have lost their expert status at the office, as everybody else can now type with word-processors. Hence, an expert system may invite the unemployment of experts but it cannot be said to be directly related to the unemployment of general clerks as the use of word-processor is not necessarily faster than the old handwriting. Machine-written documents replaced the handwritten for the purpose of increased clarity and convenience only.

Effect of Robotics and NC Machine Tools-Japanese Case

It seems that microelectronic unemployment appeared, in the Japanese case, mainly in secondary industry rather than in tertiary industry as was the case in the West. While NC (numerically controlled) machine tools were first developed in the USA in 1952, production was made much less expensive by the Japanese during the 1970s and widely employed in small factories that suffered from the shortage of skilled labour. Furthermore, around 1980, NC machine tool technology was combined with robotics, resulting not only in laboursaving but also cost reduction of products and better quality control. In 1980, more than 90% of makers and users of robotics were Japanese.[20]

European critics and labour unions were naturally apprehensive concerning the importation of Japanese robotics into their own factories and wished first to observe the Japanese experience, particularly the reaction of Japanese labour unions, of rationalization due to the advent of robots and NC machine tools in such industries as car manufacturing. Japanese union leaders as well as management may very likely not be able to provide satisfactory answers. They might argue that (1) because the Japanese economy is still growing, Japan is able to transfer any redundant labour force to other jobs, and (2) since Japanese unions are company-based, job transfer is much easier within the company than is the case with European trade unions. Large firms may be less affected by technological unemployment and redundancies could perhaps be more common in the case of smaller firms.[21]

In response to the request of European science policy makers for a report on the Japanese experience of industrial robotics, MITI for the first time responded with a report from its think tank entitled 'Investigation of possible influence of microcomputers on employment' in October 1979, as part of an OECD research project, 'The Impact of Microelectronics on Productivity and Employment'. It was submitted to an OECD special meeting in Paris the following month. Diametrically opposed to the European standpoint, MITI claimed that the introduction of microelectronics does not necessarily reduce employment and in some cases may augment employment. MITI and the Japanese government wished to evade the antagonism and resistance of labour unions and the general public, and to minimize any ill-effect on overseas sales of the Japanese computers which had been so far promoted by them.

MITI's report was described on more than one occasion as 'sheer lies.'[22] On closer scrutiny of MITI's report, one will discover that, according to the questionnaires sent to the Japanese industial sector in the early 80s, opinion was divided. As the microelectronic revolution escalates, employment in the microelectronics manufacturing industry will be increased in the sectors of design, R&D, and maintenance, with a corresponding decrease in the production sector, whereas the secondary and tertiary industries will reduce employees in all sectors. Where microelectronics can be employed for the purpose of labour-saving, it will be.

In some small machine shops in the late 1970s, manager-owners replaced skilled labor with NC machine tools. Employees in small Japanese firms are poorly unionized, and even when unionised, it is usually through company-based unions, which are much weaker than the nationwide trade-based unions in Europe.

While old skill was craft-specific, new technology is machine-specific; once a machine is altered, skilled labourers rarely adjust themselves to new NC machine tools unless they are young enough, usually less than thirty-five years old. The newly-deskilled workers are, together with the old generation machines, replaced with new generation machines and new generation labour. Such deskilled labourers simply vanish from the workforce, since their prospects for re-employment, unless the worker is employed in a large company, are virtually nil. Labourers employed in large firms at least have a chance for a job transfer within the company.

The next step: those manufacturers, who resisted the introduction of NC machine tools and retained the skilled labourers for benevolent or other reasons, are now unable to compete with those who introduced the new technology, and eventually go bankrupt.

The third step: Although no statistical study has been made on international correlation, a logical consequence will be that those European countries which fail to introduce NC machine tools and robots will not be able to compete with the Japanese export goods manufactured by NC microelectronic tools and will eventually lose their domestic market. In fact, the expansion of production and employment in the Japanese electronic equipment industry is inextricably connected with Japan's success in winning a rapidly growing slice of the world market for industrial machinery.[23] From the European point of view, this means the importation of unemployment from Japan.

In spite of sufficient warnings from scholars of the impending circumstances, the scholars were not able to provide suggestions on ways to overcome the approaching doomsday.

As a countermeasure to the seemingly thoughtless microelectronization, it could be conceivable to have a worldwide moratorium on the use of microelectronics for a limited time, to allow people all over the world the opportunity to reflect on the direction in which we are heading. Is the Luddite movement, then, as it existed during the Industrial Revolution, possible for the present-day Microelectronics Revolution? It may not be, since we, living in this technological age, may not be able to dispense with the many advantages of the microelectronic revolution.

Technology of Quality: Influence on the Third World

Robots take over the dull, dirty and dangerous jobs of cutting, welding and spray-painting in the automobile industry. While the products of old mechanical automation on a mass-production base, were certainly cheaper and of inferior quality to a craftman's work, the goods produced by NC machine tools are so well quality-controlled that they can meet consumer's demands often better than the work of craftsmen.

The goals of IC and LSI technology are not only the saving of labour, but also quality control of small-quantity production of various models. In the 60s and early 70s, in competing with the American industrial giants, such as Fairchild, which manufactured ICs through the use of the cheap labour of East and South-East Asia, Japanese manufacturers aimed for LSI fabrication through the use of computer automation and consequently enhanced the quality of products. In this case, it is not primarily labour-saving which is the main goal of this type of technology, but the pursuit of products of superior quality and precision. To this end, Japan found it advantageous to exclude human error as far as it was possible. In this way, labour-saving is merely the by-product of the main goal.

In the Third World, the effect of the microelectronic revolution has not yet been as apparent as it is in advanced countries and hence, science policy makers in the Third World still adhere to technology that is more tangible and pragmatic for solving immediate problems. In a day when the products of quality technology of microelectronics dominate the Third World market, traditional or appropriate technology of lesser quality is doomed to decline.

In the fourth and final step, the microelectronic revolution will reach the Third World, where the domestically produced goods with labour-intensive appropriate technology will be driven out of the market by the microelectronically produced goods of advanced countries, for reasons of lower price as well as better quality. Warnings of such a development have been made repeatedly by science policy scholars of the Lund Institute of Science Policy Studies since the late 1970s.[24]

There is an opinion, based on a macroeconomic view point, that technological unemployment will be eventually compensated for with technological job creation.[25] In the Japanese case, even if jobs are not created in the same sector, they will be created elsewhere as long as economic growth continues. Women high school graduates used to find jobs as bank tellers and factory workers. They are now hired in the restaurant industry, a logical development for an affluent leisure society.

This may not be the case in the Third World. Even if the relocation of labour due to the technological revolution is achieved with a minimum of social upheaval, the advent also exacerbates the differences between technologically advanced countries and the rest, since new technology is developed for and by the advanced countries. The problem has not been solved during the 1980s and will certainly become an increasingly more apparent and serious predicament in future.

Work Alienation

A traditional solution to unemployment due to the introduction of automation was to unionize labour and demand shorter working hours and wage increases. For union leaders, unemployment was a matter of quantity. They have not as yet seriously considered the quality of labour.

Management was alarmed by industrial unrest in 1972 when a wildcat strike of young labourers occurred at a GM factory in the USA. This was not for the usual reasons of wage increases and shortening of work hours, but simply because of the mundaneness of the work. Neither management nor union leaders had experience in solving this type of problem, nor could they apply traditional formulae utilized in previous labour disputes. It proved to be a jolt to management.

Since then, various countermeasures to prevent work alienation have appeared worldwide. The Swedish car manufacturer, Volvo, introduced a job rotation system which changed the kind of job every hour in an effort to reduce boredom amongst the factory workers. Japanese management initiated a QC (quality control) movement on the shop floor to encourage creative group discussions on the subject of work conditions.[26]

It has been said that the introduction of microelectronics into workplaces has also prompted changes in the quality of work. Microelectronics is described in this context as quality technology, the word 'quality' of course referring to quality as far as the product assessment of the consumer is concerned, not the working conditions of the worker. Whether microelectronics contributes toward the improvement or deterioration of the quality of work remains to be seen.

It has been argued that computerization will result in the deskilling and loss of dignity and self-confidence of workers who will become mere attendants, if not slaves, to the 'mechatronical' (word coined by the Japanese) lead player.

Technocratic apologists refute the above remark with claims that such a situation can be avoided simply through worker psychotherapy. Even so, due to the microscopic nature of microelectronics, average workers feel their work to be less tangible and less actual than previous occupations; they are merely following the given instructions, engaging in input and output without knowing what is going on in the black-box in between. This is a highly common procedure in microelectronic works, whereby, if problems occur, workers need only replace the black-box with a new one, since the complexity and cheapness of the IC render repairs economically unviable.

The introduction of microelectronics into some fields of industry increases uncomplicated programming jobs that may certainly attract and sometimes even fascinate curious young people. But generally speaking, the work is simple and monotonous. In an effort to meet labour demands for decreased work alienation, some factories abolished the Tayloristic conveyor system and replaced it with a new system in which the worker moves between machines and tasks along with the line of production so as to obtain a broad overview of production process and also, more importantly, to gain a feeling of achievement. Japanese experience of this new system shows that the system is welcome among young workers, while older workers have difficulties in adjusting.

This system also meets the labour-saving needs of management. As the number of workers on the shop floor shrinks, it makes good sense for the company to encourage a certain diversification of the tasks performed by the few who remain.[27]

As a result, a factory is virtually deserted; only a few workers are necessary to maintain a large assembly line. Human relations on the shop floor become, therefore, remote. According to a psychologist's research, the introduction of microelectronics into the workplace has resulted in increased discretional power of a worker over his machines, but a definite decrease with regard to his power in relation to superiors.[28]

Martin Trow, an American sociologist, proved that the degree of unionization is proportional to the density of population at the workplace.[29] In a work habitat such as an unmanned factory or a computer control room, the population density is close to nil; in such cases, solidarity of workers is impossible. Hence, the advent of microelectronics seems to have directly resulted in a decrease in labour union power. Thus, the 'divide and rule' form of labour personnel management is effectively at work at the new technology workplaces, whether consciously applied or not. The microelectronic revolution may cause social atomization, if not disintegration.

Quality of Work – a New-left Solution

Let us assume that it is a logical necessity that microelectronics will invite unemployment. Two classes then emerge: those who have jobs and those who do not. This is not a healthy society by any account.

It follows, then, that the limited amount of work should be equitably distributed amongst the population, resulting in shorter working hours and increased leisure time. Thus, a movement of work-sharing became a fashionable target of young people in the late 1970s, though it was criticized by the old left as governmental irresponsibility, a mere quasi-solution to unemployment.

A more fundamental issue was raised from a new-left ecologist group: that the sharp distinction between compulsory labor hours and the so-called 'leisure' hours, was not a humane or remotely satisfying way to live. It is only a post-Industrial Revolution practice to assess a worker's labour in terms of the hours constrained to machine operation, since any other method is fraught with difficulty. In other words, a worker sells his own free time to be disposed of as the employer sees fit. Such a constrained and regimented practice is not the work appropriate to human psychology and physiology and thus constitutes the major part of work alienation. In this regard, an ideal life-style may be that of the self-employed worker.

If a job is interesting and satisfying, hour constraints become less of a source of stress. It is an ecologist's task point to determine the quality of work at the sublation point between labor and leisure. No matter whether fishing is labor or leisure, it is certainly a more human work. It is the utopia Bernard Shaw envisaged 'where work is play and play is life'. In reality, this can be partially attained through a student life-style.[30]

While work alienation is progressing, there are, as their work is neither labor nor leisure, labour movements in Europe pushing for the control of the workplace by workers and the reorganization of the work process to increase interest and satisfaction amongst workers. Though their demands are partly adopted by labour management, through the practices of flexi (flexibility of work) time and the Japanese QC circles (discussion groups at the workplace), competition capitalism does not allow human ideals and considerations to be easily realized.

Privacy problems – Centralization or decentralization

It is said that the history of the electronics industry is full of the most notorious patent dispute cases. IBM, in its early history, built up a monopolistic share in the world market by blocking competitors through patents and lawsuits, causing the popular image of the computer as an enormously expensive item. IBM is culpable of having made its systems difficult and centralized, and thereby discouraging dispersed, imaginative and popular uses.[31]

On the other hand, the Japanese development of microcomputers has been aimed at the consumer market: to make computers easily accessible for everyone. One can argue that it is decentralization-oriented. We may be able to make use of microcomputers for checking against the infringement of privacy by a centralized largescale computer system.

Against the optimistic tone of the above, there is the criticism that a computer is essentially invented and developed for central control and so when it is microfied, it pervades even into private homes and infringes privacy with the use of such forms as a pocket bell and a TV telephone. Hence, one might argue that the arrival of George Orwell's *1984* is inevitable due to the advent of microcomputers.

We do not yet know whether the microcomputer is applicable for centralized or decentralized use. Perhaps it can be employed for either purpose according to the social and political context of its utilization. Is the potential exclusive utilisation of microelectronic media by centralized authorities a definite possibility? It all depends on who controls the microelectronic media in the future.

New Media Age

The terms 'new technology' and 'new media' are hardly new expressions. Whenever the industrial sector tries to find or create a new consumer market, they use the label, 'new'. But the word has lost its freshness. The new technology of microelectronics has been devalued by critical scholars, who point out its possible grave social consequences. To what extent their views have influenced the public relations departments of corporations is impossible to gauge, since the latter never explicitly admit it.

However, this does reflect a movement by corporations, in the face of the threat of decelerating demand for consumer durables, to extend their area of activity into the relatively unexplored regions of the commercial production of knowledge.[32] Their attention moved from 'new technology' to 'new media' in the early 1980s. The new media denoted a whole new set of electronic media, such as facsimile machines, TV telephone, data communication, optical communication, satellite communication and so on and is often called information technology (IT).[33] It is the problem of business profitability, rather than some impersonal force of technological evolution, which has given rise to the image of new media and its associated 'information society'.[34] The public sector joined in such a technocratic chorus, and MITI's think tank published for public relation purposes in June 1981 a fiction work entitled *A Day of Mr. S., May 21st, 1990* (Esushi no ichinichi), in which life with the new media was idealistically presented.[35] The semigovernmental Nippon Telegraph and Telephone Company (NTT) started to experiment with their new Information Network System (INS) in an experimental community in the suburb of Tokyo in the early 1980s. INS welds together the diverse forms of auditory, visual and data communications into a single web of nationwide information channels.'[36] The response of the community was not as favourable as the experimenters had hoped. Extravagant information services at home were perceived as nuisance, rather than labour-saving devices.

Mechanism of Information Capitalism

Information defined

One may define scientific work, which is openly assessed and publicly recognized, as 'knowledge', while the products of information capitalism, available only to a limited circle for private monopoly, are referred to as 'information'. In an information society, science is increasingly more identifiable with information.

Information society

New microelectronic media, glamorously advertised in the 1980s, is synonymous with another sales phrase, though slightly more academic: 'information society'. It was first coined in the late '60s by a Japanese technocrat, Yujiro Hayashi[37], who envisaged a society in which information is fully available to citizens.[38] This concept was later enlarged to mean a society that supports the general flourishment of human intellectual creativity rather than material consumption.'[39]

As new forms of media have been spread in the '80's, an information society has emerged whereby information is considered to be more important and valuable than material commodities.

If information is monopolized and kept secret by a single person or industry, it cannot be sold since the public has no knowledge of its existence. If information is publicized, it loses its commercial value, since it becomes as available to all like air. Only when information is privatized and sold to a very limited sector of the society, does it come to have a commercial value of often staggering proportions. Unlike a material commodity, information is easily duplicated and depreciated. Even if it is sold, the original owner does not 'lose' it. The above unique characteristics of information can hardly be dealt with through the traditional approach of economics, in which information was priced by reducing it to a material commodity like the copyright of a book.

Information capitalism

Tetsuo Kogawa, a young Japanese critic, coined the term 'information capitalism' in 1985[40] without interpreting its inner mechanism, merely mentioning that currency is now transformed from metal and paper materials to electronics. Recently Tessa Morris-Suzuki has elucidated on the subject of the new information society in *Beyond Computopia*, in comparison with the arguments of old-leftists, as found in P. Baran and P. Swezey's *Monopoly Capital*. The following arguments are largely inspired by her analysis.

Monopoly capitalism, in its strict sense, does not promote science and technology, but instead often buys up patents and inventions and sits on them, without putting the inventions into use, as a means of protecting its own industry from the threat of new science and technology. This may be the case in bureaucratic socialism but never so in contemporary capitalism. In contemporary capitalism, fierce inter-firm competition is the sole motivation of promotion of private science.

Private science produces information, day and night, but it is able to be copied in a short time, in spite of the pleas for the protection of intellectual property. 'In the market, only the best, that is, the newest, information is highly valuable. But since information is easily replicable, the monopolistic position will be liquidated almost immediately. Once information is sold at the marketplace, it can be easily copied; the protection of private interests of producers in this regard is technically difficult and requires tremendous extra effort. It must be treated with care, like fresh vegetables or today's catch. In spite of importunate claims made by corporations for the protection of intellectual property, legal protection covers only the immediate future; on the other hand, property soon turns obsolete. Hence, unless new information continues to be created, corporations cannot be maintained and will eventually collapse.

Scientific workers defined

Hence, management pushes their scientists and engineers to be constantly creative and highly productive. They may be more aptly termed 'scientific workers', a group which includes not only high level scientists and engineers but all those who are trained in new technology. They are a new class of people without whom information capitalism cannot be maintained. While the microelectronics revolution resulted in obsolescence and redundancy in the traditional workforce, it is this new class which is wanted by the corporate establishment to replace the old one.

We may be able to apply an old Marxist formula into the new information capitalism as follows: in the old capitalism, it was the surplus value)[41] of labour which was exploited by capital. In the case of information capitalism, it is mainly the surplus value of information that scientific workers produce that is exploited. In the so-called knowledge-intensive industry, the commodity for sale is nothing but the product of scientific workers. Corporations control not only the means of production but also the means of distribution.

In the USA, ambitious scientists start their own business ventures and sell information produced by them at an appropriate price to big corporations. In Japanese corporation laboratories, that produced by scientific workers is never reasonably priced because it is privatized and totally owned by the corporations to which the scientists belong.

Scientific workers are not particularly well paid compared to clerks and salesmen of corporations but are definitely poorly paid when compared to managers or doctors. A scientific worker's invention at a private laboratory is patented in the name of the corporation, the scientist receiving a small financial bonus[42] because the patenting procedure is costly and not affordable for an individual. The scientific workers nowadays become less and less autonomous professionals, and more and more a part of a knowledge-producing system which they are unlikely to comprehend.[43] Even if scientific workers could afford a patent, they could never put it into profitable practice. Today, a single idea or single patent does not necessarily lead to production as was the case for an old self-employed inventor. Thus, these packets of information are all channelled into the knowledge pools of corporations, which collect and accumulate the information 'goods' which scientific workers produce.

Input and output in information production

We may be able to find the most typical case of information capitalism, as defined above, in contemporary Japan in the last decade of the twentieth century. Japan has established its own paradigmatic infrastructure, while many other countries seem still to be searching for an appropriate paradigm. The Japanese infra-structure is as follows.

Information capitalism is kept soundly running as long as the input of resources, output of sales, and disposal of waste to environment are in a state of homeostasis. On the input side, most essential are the human resources of scientific workers; the process can be kept running as long as exploitable human resources are available, either through the recruitment of Japanese domestic labour or through the importation of labour from Asia and other areas. While old conventional industries are impossible without the provision of public infrastructure of harbours, road networks and electricity, gas and other utilities, the most important public infra-structure in an information society is education and training. If the public sector is unable to provide sufficient educational infrastructure, private corporations are willing to provide it themselves, as in the case of Japanese computer education which is discussed later in this chapter. On the output side, it is obvious that capitalism cannot be maintained without profitable markets in which to sell its products. Unlike the conventional commodity market, however, the potential marketability of newly produced information is unknown to users and consumers, as in the case of future generation computers. Hence, creativity of scientific workers is required to create a new market. The largest risk of the whole operation of information capitalism resides here, as an unknown market is unplannable. Furthermore, in order to market newly produced information, a considerable amount of consumer education is imperative. It is of utmost importance to analyze the structure of the industrial production of information occurring between the processes of input and output.

The structure of industrial private science

The sociology of science as an established discipline has developed considerably during the last two decades, but so far its major interest has continued to be the analysis of traditional academic science. We urgently require structural analysis of industrialized and incorporated science but, mainly because of its very corporate nature, we must confess that investigation so far is neither thorough nor penetrating.[44] Furthermore, there is no general rule applicable to all establishments which are competing with each other. Still, however, it is worthwhile to conceptualize the idealized case.

The steps of commercialized industrial science may be as follows:

1. Market demand.
2. Targeting and formulation of plans.
3. Research and development.
4. Production.
5. Advertising.
6. Dissemination into market.

Information collected at step 1 is inputted into steps 2 to 4 which are in turn subject to corporate secrecy. At the stage of basic research, classification is not yet too strict, but as the production stage is approached, extreme caution becomes more manifest. Social assessment can be applied only to steps 5 and 6 after information goods are outputted. Step 1 is then returned to, thus repeating the cycle. The surplus value created at step 3 by scientific workers, and also produced at step 4 by general workers, is privatized and invested into the next cycle again and again.

The above is still the process of the manufacturing industry, in which information value is added to the material commodity produced. In a more advanced information society, science, namely information, is not reduced to a tangible material entity but often bears its own independent status like computer software, thus curtailing step 4 of production. In an information society, the traditional workforce is increasingly being replaced by a new type of worker, namely scientific workers or information workers, from top scientists and engineers to routine programmers. They are the main producers of surplus value to be exploited by information capitalism.

In market oriented knowledge production, it would be still possible for the citizen sector to assess at step 5 and 6. More esoteric and highly privatized information is not subject to public or consumer assessment but is limited to a small circle of technocrats, thus bypassing steps 5 and 6 also.

We may schematize the structure of private science by referring to Hitoshi Yoshioka's scheme.[45]

input	process	output
public infra-structure		sales & waste
education	steps 2–4	steps 5 & 6
academic science	private information	
public science	production	
public information service		privatized information
step 1		

Shortcomings of privatized science

Taking an alarming example of computer esotericism, if one were able to gain access to satellite information on harvests throughout the world a short time earlier than anybody else, one could control the price of agricultural products as well as the stock exchange. Commercial activities in history have been and still are carried out by taking advantage of the difference of information available according to locality.

Let me take another imaginary dreadful example. Suppose a community of top space scientists exchanged esoteric preprints within their own exclusive group. Without outside pressure, they can decide the direction of the development of their research and it is not inconceivable that they could even plot a conspiracy which would drastically alter the future of mankind.

I would like to add in this connection that while the advent of printing in the modern Western world promoted objectivity and the development of academic science, contemporary microelectronic media have been developed mainly in the context of promoting industrialized, privatized science.[46] Information thus created is short-lived, soon becoming obsolete or open public knowledge. Information capitalism is kept alive in between (privatizable and marketable) information creation and waste (changing into public knowledge.) It will expand as long as the time lag between these two are increased, and shrink as it is reduced. Theoretically speaking, when the time lag ceases to exist, information capitalism, as well as the privatization of knowledge, will die out. Science as a private endeavor will also disappear. On the other hand, if the time lag is increased to the extent that citizen assessment is impossible, democracy dies out under technocratic control. When information is monopolized by a single hand, there will be no further development of private science; all a monopoly would need to do is to wait for the emergence of competitors and then crush them. Everybody is then under the surveillance of a single monopoly power – the prophecy of 1984.

SECTION II. Computer Literacy and Vocational Education

The debate on the social assessment of the microelectronic revolution in the late 70s was calmed down without finding a clear solution in the 80s. The diffusion of computers is taken for granted and now computer literacy has to become a world problem for the production of scientific workers. The Japanese public attitude to it was slow and hence the area has been dominated by the private sector. In this section, we explore the peculiar nature of Japanese computer education.

Business Initiatives of Computer Schools

Privatized knowledge which is produced becomes publicized sooner or later when its obsolescence renders it valueless. Some information is systematized, organized and arranged into a manual for users, and a text book of a new discipline, computer science, to be taught openly at educational institutions in public as well as private sectors.[47] Such institutions belong to the infrastructure of the information society.

In the late 70s, the impact of microelectronics was most deeply felt in terms of unemployment problems, which directly hit the existing workforce and resulted in job-relocation. The next problem to be raised was that of how to fit the new generation of the workforce into the new technological environment. In the 1980s, computer literacy, computer education and training became not only a matter of skill requirements for the new workforce but, in a much wider sense, the central issue of the educational sector and the society at large.

Computer literacy and computer education are, of course, enthusiastically welcomed by the private sector computer manufacturers, since obviously it is this sector which receives the foremost benefit from the diffusion of these disciplines. Such a dissemination of skills is likely to not only create a big market for their products, but also contribute to future recruitment of prospective staff. On the other hand, the education and public sectors are mostly technologically stagnant.[48]

There are four categories of computer education according to prospective use, which sales departments of computer manufacturers target:

1) makers of computers where highly professional skills are required for design and manufacturing.

2) users of computers where personnel who apply and maintain computers for their specific professional use are needed.

3) the general public to whom computer skills can be disseminated.

4) educational fields, where CAI (Computer-Aided Instruction) can be introduced into classrooms.

Senshu Gakko (Special-training Schools)

Whereas established formal educational institutions are conservative in adopting new disciplines and expanding beyond the limits of quotas, the easiest target of computer sales is the informal (or less formal) vocational education in the private sector. *Senshu gakko* are schools which specialize in a single subject of post senior-high school level. 'Konpyuta gakuin (special computer training schools)' became established in many cities, their popularity booming in the late 80s. Such educational institutions are aimed at the training of specialists of information science, experts who are needed in both the industry and service sector of the first and second categories of the above. As these personnel are required immediately, the business sector cannot be expected to wait until public sector education incorporates such schemes into their curriculum.

The special-training school system started in 1976 and has grown rapidly since then.[49] The special-training schools are vocational schools, comparable to the American community college network, which give one-year or longer schooling on specific vocational subjects, such as information processing, accounting and so on. The special-training schools flourished perhaps partly because Japanese universities are institutions of labelling, designating student 'quality' according to their strict entrance examinations rather than real education and training. Hence, those who seek substantial training to cater for their immediate educational needs are attracted by the utilitarian merit of specialtraining schools.

Computer schools of this special-training school category attract not only young schoolleavers (senior high school graduates who do not enter colleges or universities) but adults sent by their firms for job transfer and college students who are not satisfied with formal university curricula. More than half the total of information science students (35,000 out of a total 66,000 in 1988) in Japanese higher education are trained in these informal specialtraining schools.[50] While OJT (on the job training) is carried out in inhouse schools of big corporations, Off-JT (off the job training) for retraining of corporate employees, is conducted by 80.6% of enterprises (1987) by sending them to computer schools etc. Jumping on the bandwagon of the success of special-training schools, many private universities plan to open new schools and departments in information science, a move which has been subsequently chartered by the Ministry of Education, Science and Culture (MESC, formerly Ministry of Education).

While these informal special-training schools are chartered not by the MESC but by local governments, the MESC has supported the schools since 1983 with subsidies for the purchase of computers and large scale educational equipment.[51]

Despite the growth during the 1970s of the number of computer science departments from 7 to 43[52] in Japanese universities nationwide, the equipment and facilities are far inferior to those of the private sector, because of the impoverished nature of the Japanese academic sector.

The next target of computer sales departments is a nebulous category, the third category mentioned above, namely, a new nationwide system of 'life-long education' or 'recurrent education'. The public sector, particularly local and district governments, is interested in educating citizens in the technology of information facilities to a minimal level of proficiency, without any in-depth knowledge, at such local centres as science centres, culture centres, libraries and museums. Such training is aimed at prospective users in the future expanded market.

In spite of the corporate strategy and effort described above, sales have not penetrated the hard core of the education establishment, the last resort of the last category. Education is a marginal sector of expansion in the microcomputer market, compared to businesses and the home.[53] This is a world phenomenon but especially so in Japan, as compared with the case in the USA, perhaps because of the inflexible bureaucratic nature of the Japanese educational system.

The Response of the Education Establishment

Since the educational field is undoubtedly one of the biggest markets for microelectronic equipment, sales departments of corporations aimed for the 'mechatronization of education', by applying a similar strategy to that successfully adopted for 'mechatronization of health care and hospitals' during the 80s. The latter created a new machine-intensive system of health care led by medical equipment manufacturers rather than the initiatives of doctors and patients.

Education is traditionally very labour-intensive work. The introduction of the microelectronic revolution into a classroom will certainly promote unemployment of teachers, if it is utilised for the purpose of labor-saving in the teaching profession. Like other unions, a teachers union would fiercely fight such a move.

The ability of teachers is measured and assessed in terms of the quality of work rather than efficiency and productivity of commodity manufacturing. Hence, a teachers' union has a better cause for opposing the advent of computers compared to other unions. Japan has a precedent case in the recent past in which the introduction of TV teaching at a classroom was vehemently resisted by teachers in the 1950s. A cynical Japanese critic commented at the time that the Japanese educational practice is so formal that teachers can be easily replaced by computers. Hence, teachers' opposition to the introduction of computers in classrooms.

Yet, the provisions newly designed as a response to technological challenge are increasing the complexity and diversity of the curriculum, adding demands on the teachers, making their working agendas heavier still.[54] Unless early R&D for computer education is geared to labour-saving, technological unemployment may not occur in the education field. Hopefully the introduction of computers will liberate teachers from formality and result in the fostering of humane qualities in the educational community, just as the advent of photography made an impact on the world of art in the nineteenth century. Some young ambitious teachers are utilizing the new technologies and developing new educational software, and seem to be propelling education into a more machine-intensive as well as labour-intensive 'high-tech, high-touch' direction.

Apart from immediately relevant unemployment cases, the education establishment is powerful enough to resist policies that ignore liberal values in preference to vocational values and respond only to economical and technological change. The educational sector calls for a broad vocational foundation for all workers, though many employers prefer short and limited training focused upon particular job assignments. Moreover, the problem of outdated equipment and instruction-one already familiar to vocational education-is now magnified with the advent of computers.[55]

The Reaction of the Public Sector

In the USA, the early phase of the process of knowledge production is sponsored by the public sector, including local as well as state governments, with the slogan of 'computer literacy', perhaps in order to outstrip Japan in the future computer war. In contrast, to the surprise of Western observers, the speed of diffusion of computer literacy in Japan is the lowest among OECD member countries.[56]

The Japanese government, the Ministry of Education, Science and Culture (MESC), is still reluctant to sponsor computer education in the public sector, leaving it to inhouse training of the corporations or private vocational computer schools. The MESC explains that this is due to the fact that Japan does not have such monopolistic computer manufacturers as American IBM and Apple, and so the government cannot decide which company's hardware to adopt as equipment to be installed in the public school network nationwide. The government also claims that it is unwise to adopt a model at present due to the unstable state of affairs of current computer developments. They argue that, with model changes every half-year, a machine-specific training will be obsolete by the time a student graduates. Even if computer education is adopted at schools, nearly half of the student cohort would not take it seriously unless it could be geared to the university entrance examination. Since this is not the case, students do not normally allocate time for it, other than for personal interest.

In practice, most Japanese college students graduate without having any chance to learn computers during their formal education. CAI is only extensively used in the national technical college system (three years of senior-high school standing plus two years of junior college level), where students are freed from preparation for college entrance examinations.

Only at vocational senior high schools of post-compulsory education level (ninth to twelfth school grade) where students are not expected to enter tertiary education, are computer courses adopted and word processors used in the place of old abacus training. In the 80s, courses on information engineering and information processing are introduced into commercial and technological vocation senior high schools to provide lowerclass information engineers.[57]

The public sector should not partake in the pursuit of private profit, but maintain not an economic, but a development rationale. Accordingly the building of the infrastructure of the microelectronic revolution must be accomplished by the government, just as MITI protected and encouraged the computer-manufacturing sector. More important still, it is the function of a government to regulate and resolve the conflict between different companies and possibly bring about uniformity in computer hardware as well as software, since it is the consumercustomer, not industry, which suffers from the inter-firm computer war. This is rather close to the policy stance of the Ministry of Education that dominates Japanese education, while MITI (Ministry of International Trade and Industry) has so far encouraged inter-firm competition under its control.

From Bureaucracy to a Business Model of Education?

Modern Japanese education was planned and organized by a bureaucracy to act as a department of bureaucracy in the 1870s. The bureaucratic model of education has the following characteristics as opposed to the private model:

bureaucratic	business
1. official character	privatization
2. high regimentation	dexterity
3. formality	pragmatism
4. inflexibility	flexibility
5. inefficiency	efficiency
6. long-range view	short-sightedness
7. accountability	profit-making
8. stability	changeability
9. stasis	dynamism
10. regulation-based	money-based

While most Japanese formal education belongs to the bureaucratic model, special-training schools and recurrent education more closely resemble the business model.

The typical bureaucratic model of education is found in the civil service examination system of traditional China, where education was subordinate to bureaucracy. The modern Japanese education system was created within the existing bureaucracy in the 1870s, and the government adopted the civil service examination in the early 1890s. In this same period, the spirit of talent recruitment by way of fair examinations was all over Europe. Japan followed suit. Japanese examinationmindedness may be attributed to the culturally embedded Confucian spirit that is shared in Confucian cultural regions, such as Korea, Hongkong, Singapore, Taiwan, and, of course, mainland China. It is the spirit of the civilian control over military by scholar bureaucrats, who are selected rigorously and fairly in terms of pure talent shown on examination sheets, thus excluding the nepotism and pressure of the wealthy and powerful.

The logical consequence of a Confucian meritocracy is the creation of a 'gakureki shakai' (a society hierarchically ranked according to the quality of school attended). In its simplistic form, all Japanese people are hierarchically ranked, from the graduates of the University of Tokyo on top of the pyramid, to the compulsory school leavers at the bottom. When two people meet for the first time in such a society, one cannot feel confident and comfortable until one knows what school the other has graduated from.

In such a society, the social position and status of each person is predetermined at the time when one enters the school at which one will finish one's education. It is a static and stable society, and, since all are content with their social rank, as fairly and equitably assessed by examinations, law and order is well maintained.

The above is, of course, a much idealized form of a bureaucratic model society, in which only the ruling top elite could be one hundred percent happy. The remainder of the population would attempt to climb up the social ladder, the most effective means being through business success.

The Balance Between Public and Private Education

Japanese modern history has been a constant process of technology transfer from the public to the private sector,[58] the process being completed only in the postwar period. Compared to technology, education has been a much delayed component of the public-private transfer. Whether education should be completely privatized is another question, however.

The balance between public and private expenditure on the investment of computer education varies from one country to another. Recent European trends show that the public sector, particularly in France, is the largest promoter of computerization of education.[59] Since Europe, in the 1980's, was the late-comer in computer manufacturing, the government must push their education community as the market for their industry. In the developing stages of Europe's computer manufacturing industry, the public cost may be accounted for by the benefits of the nurturing of a future industry and job-creation. In the Japanese case, the private cost will be amply returned. The Japanese private sector is so all powerful that it does not require government assistance.

We said in the beginning that a new technology is promoted by technocracy and assessed by democracy. In reality, however, new technology is promoted by the private sector and assessed by the public sector. It is, therefore, the essential role and responsibility of the public sector to control and regulate the computerization of education, especially at a time of ruthless cutthroat competition when long range consequences are not on the agenda for the private sector competitors.

Intrinsic Problems of Computer Teaching

The Teachability of a New Technology

It is traditional educational practice that adults are taught to maturity by adults. In a new technological environment, however, adults are required to maintain the childlike curiosity and childhood ability to learn and work with new technology.[60]

Whether new technologies can be teachable is an intrinsic question of technical education. Nobody is qualified to teach, at least within any formal education system, the *enfants terribles* who manage to break through to the technological forefront. For instance, the creators of the Apple personal computer, Stephen Wozniak and Steven Jobs, share many common characteristics; both were bored by school and dropped out of college to work in the niches of electronic hobbyists.[61] Teachers in formal education in Japan cannot compete with the computer skills of teenage hobbyists who frequent the Akihabara area (the famous centre of micro-electronics sales in Tokyo).

When the frontier is still moving ahead, those involved have no time to systematize the accumulated knowledge into a teachable form and often they are confused as to which is the right track to develop and which is not. At this stage, if the technology is teachable, it can be only through onthejob apprenticeship. Only after the forefront of computer developments become stabilized and knowledge standardized, can the subject truly becomes teachable. Before IBM introduced its own personal computer, the IBM PC, in 1981, teachers had no standard computer hardware for teaching in vocational education. Only when the subject has become solidly established to the extent that a textbook can be written, can it become a school discipline to be taught in formal education.

Vicissitudes of Discipline

According to Keiichi Yamada and his colleagues[62], the average lifecycle of science and technology is seven years from the birth of a paradigm to the disappearance of a research front. The length of the lifecycle has dramatically shortened in recent years. In the past, an engineer who mastered a technology at school could earn a living from his skills for his whole career. Nowadays, because of high competition, the lifecycle is so shortened that technicians and engineers must be retrained two or three times to fit themselves to the new lifecycle of new technologies. Those who fail are obliged to retire. The lifecycle can be applied to top ranking scientists and engineers also. In order to meet the demand for polymer chemistry in the postwar years, the University of Tokyo established the Department of Polymer Chemistry within the School of Engineering, but when it was founded and students graduated, the high time of polymer chemistry was already over.

In science and technology, such rapid change can be observed in almost any field. In terms of research performance, the lifecyle of research and development can be clearly recognized. In terms of manpower, it is like a business cycle of supply and demand; students are attracted in an area in which there exists a certain manpower shortage at the time of college entrance, and when they finish training, the area is overcrowded with those who entered in the field with the same motivation. Furthermore, the vicissitude is accelerated with the change of science policy in the public sector. It is still an unforgettable and sour experience of the American scientific community that they have suffered from severe budget cuts since 1968.

Unlike USA, France and UK, and like West Germany, the Japanese budgetary structure does not allow dramatic increases or decreases in budget allocation, but rather a slow and gradual incrementation, intensified only in the 80s. This may be perhaps due to the absence of a definite science policy in Japanese government. A Japanese system of entrance quota for colleges, has also contributed in stabilizing the vicissitudes of disciplines. Since the system is conservative, Japanese formal technical education does not flexibly meet the demands of the industrial sector, but is ironically less vulnerable at times of cyclical employment crises.

The departments of electronic technology in Japanese tertiary institutions have prospered since their emergence around 1960. If we look closely, however, they have undergone many changes. Parallel to the lifecycle of successive generations of electronic technologies, from the vacuum tube, the transistor, integrated circuits, largescale integrated circuits (LSI) and finally to VLSIs, scientists and engineers have experienced harsh years of retraining and readjusting themselves to the new generation of technology.

Is Curriculum Reform Possible?

New technologies create new disciplines and demand curriculum reform at educational institutions. This is particularly noticeable in the teaching of science and technology, for obvious reasons. Even if a new discipline is successfully introduced into the curriculum, old disciplines and old teachers never die out. Consequently, the curriculum becomes overcrowded and the students' burden increased. Only in the event of drastic institutional reform is it likely that a radical shakeup of the overcrowded curriculum would take place. The campus unrest in the late 1960s provided one of such rare opportunities.

A logical and consistent attempt at such curriculum reforms was projected under the title of 'Engineering Science' in the 1960s. The advocates of the project rearranged all the disciplines, putting undisciplinized knowledge and even newly compiled empirical findings under a thorough methodology, abstracting common features out of a variety of phenomena, simplifying the data into minimum essentials following the logical steps from the fundamental to the applied. This approach can be also applicable to other scientific areas such as the medical sciences, where the overcrowded curriculum is a serious source of discomfort.

In reality, however, the Engineering Science project has been disseminated only among a small circle of advocates and sympathizers. It is nearly impossible for old teachers to be converted into a new teaching method. Thorough reform must be postponed until the time of the next generation.

The teaching of drawing, for example, has been the basic of all modern engineering vocation since the time of the French Revolution. With the advent of computers, traditional teaching of drawing would be largely replaced by cadcam teaching and computer science. The old teachers of drawing who belonged to the precomputer generation are totally unqualified for the teaching of computer science. Such a state of affairs is similarly occurring in any field that has introduced computers and will continue to occur in the future as well.[63]

A Choice for the Next Generation?

When a new technology is introduced into a society, the immediate reaction of the average people who are content with their customary lifestyle is embarrrassment and subsequent rejection. The history of technology has ample examples of such cases as the automobile, the aeroplane and so on. A new technology thus introduced, divides mankind into two categories, those who take advantage of it and those who do not. In cultural terms, they can be divided into those who can adjust themselves to the computopia and those who cannot; namely, computer literates and illiterates. Or, in a political sense, the two camps of technocracy and democracy. The final resolution of this awaits the next generation. What will be their choice?

Notes

1 Gregory, *Japanese electronics technology*, p.6.I
2 *Denshi kogyo nenkan* (Yearbook of the electronics industry, Denpa Shinbunsha,1979). Sachio Yoshimura, *IC sangyo daisenso* (IC industrial warfare, Daiamondo, 1979).
3 Morris-Suzuki, *Beyond Computopia* pp.29–30.
4 Yoshiro Minamizawa, *Nihon konpyuta hattatsushi* (History of computer R & D in Japan, Nihon Keizai shinbun, 1978) p.180ff.
5 The idea of research associations was introduced into Japan in the early postwar period through the Japanese translation in 1951 of J. D. Bernal *Social Functions of Science*, which was originally published in 1939. It was widely studied among concerned Japanese scientists who were responsible for reorganizing research organization in the postwar period.
6 Semi-governmental agencies such as *Denshi Shinko Kyokai* (Association for Microelectronics Promotion) played a major role in this function.
7 Gregory, *Japanese Electronic Technology* pp.50–51.
8 Iann Barron & Ray Curnow, *The Future with Microelectronics* (Frances Printer, 1979).
9 Counter Information Service, *The New Technology* (1980).
10 Richard Layard and Stephen Nickell, 'Unemployment in Britain', in C. Bean, R. Layard and S. Nickell (eds.), *The Rise in Unemployment*, (Basil Blackwell, 1987), p.166.
11 Simon Nora et Alain Minc, *L'Information de la Societe* (La Documentation Francaise, Paris, 1978).

12 Tom Forester (ed.), *The Microelectronics Revolution*, (Basil Blackwell, 1980), p.303.
13 Calvin C. Gotlieb, 'Computer A Gift of Fire' in S. H. Lavington (ed.) *Information Processing 80* (North Holland, 1980).
14 An earlier one is Freeman, C. M. 'Unemployment and Government' J. D. Bernal Memorial Lecture, May 1978 in T. Forester (ed.), *The Microelectronics Revolution* (Blackwell, 1980) p.308 ff. C.M. Freeman, J.A. Clark, and L. Soete, *Unemployment and Technical Innovation: A study of long waves and economic development*, (Frances Printer, 1982). And also C.M. Freeman, and L. Soete, *Technical Change and Full Employment (1987, Basil Blackwell). Donald Leach and Howard Wagstaff, Future Employment and Technological Change* (Kogan, 1986).
15 *Bunka toshiteno sentan gijutsu* (High technology as culture, NHK, 1985) p.81.
16 Gregory, Gene, *Japanese Electronics Technology: Enterprise and Innovation* (John Wiley, 1986), p.45.
17 *Ibid.* p.408.'
18 CEDEHOP (European Centre for the Development of Vocational Training) *Artificial Intelligence and its Potential as an Aid to Vocational Training and Education*,(1988, Office for Official Publications of the European Communities), p.17.
19 *Ibid.* p.16.
20 The figure depends on how one defines 'robots'. There is another conservative estimate of 70 %.
21 Tessa Morris-Suzuki, *Beyond Computopia: Information, Automation and Democracy in Japan*, (Kegan Paul International, 1988) p.93.
22 Kei Tsugawa 'Conpyutaka wa nani o motarashita ka (What has computerization brought about?)' *Gijutsu to ningen* July 1980, p.129.
23 Morris-Suzuki, *Beyond Computopia* p.97.
24 *The Lund Letter on Science, Technology and Basic Human Needs* no.12 (1979).
25 Stephen G. Peitchinis, *Computer Technology and Employment Retrospect and Prospect* (Macmillan, 1984).
26 Shigeru Nakayama, 'Maikon kakumei no shogeki' ('Impact of Microelectronic Revolution'), *Ekonomisuto*, 10 April 1981, reprinted in Shigeru Nakayama *Shimin no tame no kagakuron* (Science for Citizens, 1984) p.94. A leader of the QC movement at the shop floor told me that the new machinecentred workplace is so dehumanized that one cannot withstand it psychologically without QC circles.
27 This is exceptionally true in the case of Japanese management. It seems that in some other countries there has been more reluctance to integrate operations in the hands of a single worker. Morris-Suzuki, *Beyond Computopia* p.115–116.

28 A Zimbalist, 'Worker Control over Technology', *The Nation* Nov.17, 1979.

29 Martin Trow

30 Shigeru Nakayama, 'Honrai no gakushu to shigoto nitsuite' ('On true learning and work'), *IDE* January, 1979 and also his 'Mirai no shigoto to ekoroji' ('Future work and ecology'), in *Shimin no tame no kagakuron*, pp.107ff.

31 Theodor H. Nelson, *The Home Computer Revolution* (1977) pp.42–43.

32 Morris-Suzuki, *Beyond Computopia* p.61.

33 Christopher Freeman and Luc Soete, *Technical Change and Full Employment* (Basil Blackwell, 1987), p.51.

34 Morris-Suzuki, *Beyond Computopia*, p.67.

35 Hitoshi Yoshioka , *Tekunotopia o koete* (Beyond Technotopia, 1982) p.197.

36 *Ibid.*, p.33.

37 The original Japanese is 'johoka shakai'.

38 Yujiro Hayashi, 'Kogyo shakai no yukue' ('Future of industrialized society'), in Nakayama (ed.), *Nihon no gijutsuryoku* pp.166–168.

39 Yoneji Masuda *The Information Society as Postindustrial Society* (Institute for the Information Society, 1980) p.3.

40 Tetsuo Kogawa, *Joho shihonshugi hihan* (A critique of information capitalism, (Chikuma, 1985).

41 Here, 'surplus value' is employed not in a rigorous quantifiable fashion but merely rhetorically, as the 'surplus value of information' can never be quantified.

42 The bonus is extremely small and usually equivalent to several tens of US dollars.

43 *Beyond Computopia* p.123.

44 A questionnaire based study was recently conducted. Shogo Sakakura, 'Minkan kenkyusho o chushin to suru kenkyu kanri no jittai chosa hokoku' ('Report on current research management in private sector research institutions'), *J. Science Policy and Research Management* vol.2, no.3 (1987).

45 Hitoshi Yoshioka, *Kagaku kakumei no seijigaku* (Politics of scientific revolutions, Chuo Koron, 1987), p.117.

46 Shigeru Nakayama, 'The three stage development of knowledge and the media', *Historia scientiarum*, no.31 (1986) pp.101–113, and also 'Privatization of scientific information', *Todai Symposium, '86: Information and its functions*, Institute of Journalism and Communication Studies, The University of Tokyo (1988).

47 Shigeru Nakayama, 'Paradaimu, deishipurin, kyokasho' ('Paradigm, Discipline and Textbook'), *IDE*, December, 1982, reprinted in *Shimin no tame no kagakuron* pp.29ff.

48 OECD, *New Technologies in the 1990s A Socioeconomic Strategy* (1988) p.110.
49 Table 'The growth of *senshu gakko', from Seishonen hakusho* (White paper on the youth, 1988) p.362., *Wagakuni no bunkyo shisaku* p.42.
50 Monbusho (MESC) ed., *Wagakuni no bunkyo shisaku* (Japanese education policy, white paper, 1988) p.447.
51 *Ibid.* pp.283–284.
52 *New Information Technologies*, p.77.
53 OECD, *New Information Technologies a Challenge for Education* (1986) p.14.
54 *New Information Technologies*, p.100.
55 New Technologies in the 1990s pp.102–103.
56 *New Information Technologies*, p.20.
57 Figures are from the table on Monbusho, *Wagakuni no bunkyo shisaku* p.447.
58 Shigeru Nakayama, 'Science and Technology in Modern Japanese Development', William Beranek Jr. and Gustav Ranis (eds), *Science, Technology and Economic Development: a Historical and Comparative Study* (Praeger, 1978) pp.202–232.
59 *New Information Technologies* chapter V.
61 Stan Augarten, *Bit by Bit: An Illustrated History of Computers* (Unwil Paperbacks, 1984) pp.276–277.
62 Yujiro Hayashi and Keiichi Yamada, *Kagaku no raifusaikuru* (Lifecycles in science, Chuo koron, 1975) p.48.
63 Shigeru Nakayama , 'Gijutsu kakushin to kyoiku' ('Technological Innovation and Education'), *Shin kyoiku jiten*.

that had prevailed amongst the population in the 70s. They addressed this slogan only to the Japanese people. It is unlikely that they could have anticipated the wave of international reaction against it.

I have translated this word as 'technonationalism', a term which would certainly evoke the feeling of threat among the neighbouring countries which were occupied by Japanese military forces during the war. As it is a universally held postwar image that science and technology are indispensably rooted in military build up, the slogan could be interpreted as the renewed military aggression of Japan and create a new source of misunderstanding of the Japanese. Thinking of science and technology in terms of nation states is the problem that surrounds the basic characteristic of modern science and technology. In this way, instead of aligning with peace and democracy, science and technology may align with dictatorship and authoritarianism. The Iranian revolution that overthrew the Pahlavi regime and the anti-establishment movements in South Korea both demonstrated a kind of understanding in which science and technology are equated with the 'technocratic system' in which science and technology came to be understood as primarily antagonistic and suppressive toward the citizenry.

In order to warn of the potential grave consequences of their naive technological optimism, I wrote a feature article in a magazine on technonationalism[4]. To what extent my efforts influenced Japanese science policy makers, I do not know. At least those with whom I have directly spoken became concerned with the prejudicial tone of technonationalism and avoided using the slogan of 'gijutsu-rikkoku' again.

The rationale of technonationalists tends to run like this: 'Japan is a country scarce in natural-resources. To counter this state of affairs, Japan has to be strong in other areas; it must be science and technology'. This logic was used repeatedly during the war to rationalise the military budget as 'we have to build up a strong military to compensate for our scarcity of natural resources, such as petroleum and rubber'. Such hypotheses were again revived in the postwar period, replacing military effort with the imperative of export and then the imperative of Japan's science and technology. It has developed virtually into a national psychosis: the scarcity of natural resources is implanted in Japanese minds from childhood, and is often used as a means of national mobilization.

Ecologists may disagree about whether a scarcity of resources is a necessary and inevitable outcome of industrialization; nevertheless, Japan still possesses plenty of water and forest resources compared to most other countries, which can be preserved at the cost of the natural resources of surrounding countries[5] which are imported into Japan.

In the meantime, Japan has been involved in science and technology conflict on the matter of intellectual property ownership with the US. The Japanese sought a way to settle the hostilities of the USA and, to this end, adopted the new slogan of 'internationalization' as a buzz word, instead of 'technonationalism', which is now used by Japanese critics to describe those advocates behind America's claims of unfair trade practices. This changeover of slogan implies a general switch from competition to cooperation.

Plea for Creativity

It has long been said as a criticism of Japanese science and technology that it is mere imitation, devoid of creativity. Speaking from the point of view of the academic sector, criticism originates from the universal criteria of assessment by the world scientific community and therefore it is a question of pride in the Japanese scientific community, not to mention national pride. It has, however, no direct relationship with economic value. Hence, it has not attracted the attention of the industrial sector.

Whenever the wave of international liberalization (of trade, capital and technology) presses on the Japanese postwar industrial community, there are always those who fear a complete sweep over of markets by foreign enterprises. Leading industrialists then follow with calls for greater support of Japanese science and technology in order that it may compete with foreign science and technology. Often the support is not forthcoming; up to the 1970s, it was the general belief of the management camp of the industrial sector that it was cheaper and quicker to acquire technology from the US and commercialize it, than to spend money on research and development in Japan.

After entering the 1980s, however, when the US began to think of Japan in adversarial terms as a technological competitor, the Japanese industrial sector came to the realization that the acquisition of technology from external sources was problematic and that it would be impossible to maintain competitiveness without promoting its own creativity. In other words, the industrial sector came to recognize for the first time in the mid 1980's the value of creativity over imitation. As the value system changes in Japanese private science, the whole community shall move towards that new end. Although to what extent this can be attained will remain to be seen.

Greater technical innovation was also advocated by external groups. In the mid-80s, there appeared a number of American critics who claimed that contemporary Japanese science and technology was weak in the basic sciences, and that it lacked creativity[6] and hence would not progress in the long run.[7] These comments were by no means new in the self-criticism of the Japanese scientific community, but what was new was American concern with their new hypothetical enemy, Japan. Behind these comments, one can sense the defence of American intellectual property based on their criticism that the Japanese spend much less on basic science (namely, academic science as practised in universities), borrow the fruit of American basic science and commercialize it without paying the basic cost.

Creativity in Academic Science

The Ministry of Education has been, throughout the postwar period, the supporter of university research and academic science; their target has been, in a simplistic expression, aiming toward the production of more Japanese Nobel laureates, in contrast to MITI's support of private science. The Japanese academic sector, backed by the Ministry of Education (Monbusho), took advantage of this outer criticism and tried to effect an increase in the basic research budget for academia, though so far without any marked success[8] Oshima defended Japanese spending on basic research, remarking that Japanese industry spent higher percentages of research expenditure on basic research than Western countries. Further, the percentage of all expenditure on basic research spent at the Japanese universities is higher than those of Western countries, in Keichi Oshima, 'Research and Development and Economic Growth in Japan' in B.R.Williams ed. *Science and Technology in Economic Growth* (Macmillan, 1973) p.318. This argument depends, of course, on the definition of basic science employed in each country.

If we assume university research as the site of basic research, then the creativity of Japanese universities compared with those of the US is certainly low, but that of Japanese private science as practised in private laboratories is relatively high.

The Japanese university system is a closed institution and hence universitybased academic science is much less competitive than that of the American system. Thus Japanese academic science is inappropriately low for the present-day status of Japanese science and technology in the world community. Certainly, the number of Japanese scientific papers and their share in the total database of world scientific community is increasing as the contribution of industrial laboratories increase.[9] Neverthless, the above features may not be easily altered. Japanese businessmen and officials at MITI have given up battling with MESC, which maintains a tight grip on the nation's educational system. 'MESC is hopelessly old-fashioned,' they sigh.[10] Hopefully, the rich private sector will be willing to inject funds into training and research in the nonuniversity sector. This move may not be without repercussions for university research.

Creativity in Private Science

Pleas for creativity in Japanese technocracy appeared repeatedly in the media throughout the 1980s. Some attribute the lack of creativity to the Japanese national character and some to the educational and research system. Without analyzing the mechanism of different kinds of scientific endeavour, however, such arguments remain at a superficial level and lead, invariably, nowhere. The superficiality of the national character argument is rooted in the assumption that the gift of scientific temperament is variable according to the race. The present writer does not take this argument seriously, as the migration of students can prove otherwise; it is safe to assume the equal distribution of talent irrespective of race. The argument attributed to institutional differences is also superficial as it is largely derived from the failure to recognize the difference of creativity in academic and private science, and hence, from a misleading argument which applies the criteria of the former over the latter.

Basic science and innovation, as conceived by the private sector, aims for the creation of the seeds for future commercialization, namely 'mission-oriented basic research', which differs considerably from basic science in the academic sector. In other words, creativity in academic science and that in private science are two entirely different things.

A creative idea takes shape in the mind of an individual, though quite often it is spurred on by discussions with others. Academic science in prewar times was basically an individualistic activity, in which all stages of research, from the initial generation of ideas to experimentation and to the final step of publication, were carried out by a single person. Although it is still so in many areas of academic science, a division of labour has been necessary in the capital and resource intensive big science projects of the postwar period, in which the merit of individual creativity is hardly recognised as significant or important.

In private science, the situation is far more complicated. A craftman's art or a self-employed inventor's patent may be put into use by an individual but modern private science can only be meaningfully employed when all steps, from idea generation, research and development, production, advertising and sales, are successfully completed. These steps are not achieved by a single hand.

We are familiar with the big names of achievers in the chronological table of the history of science and technology, such as Albert Einstein and James Watt. In the postwar period, however, a historian of science experiences increasing difficulty in naming a limited number of present-day heroes in successful scientific achievements, especially for a big science project in which the authors of an article can reach a two-digit number. The best established criterion of scientific creativity, the Nobel prize, can only be applied to the traditional small sciences as there is a limit of three laureates for a given topic.

In private science, the situation is still more difficult. The invention of transistors still can be attributed to Schockley but it cannot be done in the same way for the Japanese VLSI project, which produced 1,000 patents during its first four years.

If creativity is only a question of having an ingenious idea, the plea for creativity will be fulfilled by eliminating the conditions which prevent the generation of such ideas. For academic science, this will still be a worthwhile prerequisite for the promotion of academic science.

For private science, a creative idea is undoubtedly the necessary condition but it alone cannot accomplish a thing, but will instead simply vanish into oblivion. A technological achievement is conceivable only at the place where provisions are made for not only necessary conditions but also sufficient conditions, the infrastructure to turn an idea to reality. If a piecemeal ingenious idea is everything, a corporation does not need its research laboratories. There is a classic example worth citing here. The principle of atomic fission was discovered by Otto Hahn, a German physicist, but an atomic bomb was not manufactured in Nazi Germany. Only when the seed was transplanted into the American infrastructure, did the historic technological breakthrough come into being.

Sheridan Tatsuno has explained this 'infrastructure' rather nicely:

> America's romanticized notion of the rugged individual, the lone scientist/inventor like Albert Einstein, or Apple Computer's Steve Wozniak, is giving way to the more pragmatic reality of large, corporate R&D teams conducting joint research. As American researchers have long known, unbridled individualism alone does not guarantee creative research. Adequate funding, supportive management, excellent laboratories and equipment, 'shoptalk,' seminars and conferences, teamwork, and a sense of purpose are the crucial ingredients for success.[11]

The Structure of Imitation

Japanese imitation has often been criticized by foreigners and the Japanese themselves. Such criticisms are, again, made with academic science in mind, a social sector in which imitation is dishonourable and often considered to be an infringement of the professional ethics of the scientific community.

In private science, the situation is, again, complicated. In the country from which a technology originates, the normal procedure of implementation involves the generation of an idea, R&D, testing, production and, finally, marketing. In a developing country where an advanced technology is transferred, the step of the catching up of technology is reversed. That is, people in developing countries start out at the last step of sales, at the right end of the following spectra, gradually following up the original developmental steps backward from right to left on the downstair spectrum. By catching up in all aspects of the technological infrastructure, they finally attain independence and autonomy in technology. At each stage, different ways of technology transfer should be considered.

Developed: idea R&D model test production sales

Developing: idea R&D model test production sales

In academic science, a number of cases have been reported that a novel idea was not adequately assessed by the discoveror's peers who were not able to evaluate its revolutionary value at the time because it was not the central issue of the contemporary scientific community. Most of them are not even recorded but simply disappear. In private science, it is rather rare but there are examples of happy cases in which a novel idea is implemented into practice and survives. Because of the privatized nature of private science, some patented ideas were bought up and neglected by corporations whose existing technology and products might be jeopardized in the market if the ideas were put into practice.

In the case of developing countries, creativity does not mean much for the immediate goal of catching up with advanced technology. Even if scientists in the developing countries come up with an ingenious idea or a remarkable finding in research and development, they lack the infrastructure to put the ideas into practice to the final stage of production. Such originality and creativity will be totally aborted and unable to develop. The history of science must be full of aborted ideas but very few are recorded.

In academic science, both the performers and peer assessers share the same value system but in private science, the researchers do not necessarily share the same problematique with their sponsoring assessers. Thus, it is ineffectual to merely call for creativity and apply psychological pressure on industrial researchers to perform.

The problem of creativity in private science is mostly the matter of assessors rather than that of performers. It is the role and responsibility of the assessor to put creative ideas into practice; otherwise, creativity at the idea stage disappears without being linked to 'technological innovation', which is quite a different thing from creative ideas. The Japanese private sector has so far not needed or has not attempted to take a risk on the creativity of its scientists; Japanese industry has been content to recruit college graduates with high examination records, normal scientists, rather than ambitious non-conformists.

Exceptionally lucky cases arise when the idea is saved through its being transplanted into an appropriate infrastructure, and subsequently materialises. There is evidence of early Japanese ideas and research which were eventually put into practice in the USA. One of the most notable cases is that of Reona Esaki, a Japanese Nobel laureate in physics, who was brain-drained from Sony into IBM's Watson Laboratory. Recently cases tend to occur vice versa. The high electron mobility transistor (HEMT), which springs originally from an idea proposed in 1969 by Esaki and others at IBM's laboratory, was pushed forward by Fujitsu to perfect the super-high-speed computer.[12] Whereas technological infrastructure is tightly tied up with local cultural, economic and other conditions, scientifically creative ideas often possess universal applicability, and hence transplantability. Such a breakthrough idea is visible and often publicized; hence it is able to be stolen.

The history of Japanese private science shows abundant cases in which students who were sent abroad to study the universal aspects of Western science through Western higher education, returning and spending their entire lives transplanting and implementing the basic Western ideas, namely the build-up of an infrastructure appropriate to local conditions. This is nothing but a Japanese version of 'appropriate technology' which was in vogue during the 1970s.

Once that all levels of the infrastructure have been established, creative ideas would be directly put into their own infrastructure and implemented. I would like to add that creativity is required, in such a complicated system of private science, at all levels of the spectrum, R & D, design, production, advertising, market research, and sales. Such potential for creativity is not limited by what occurs at the very first stage of idea generation. The above historical experience of Japan could be followed by most Asian and South-Asian countries.

Internationalization – Emphasis on Cooperative Phase

'Internationalization', employed as a government slogan, is another aspect of Japanese technonationalism. As exemplified by the IBM industrial espionage incident in 1982, which was reported sensationally, in the mid-80s, the Japanese technocratic establishment was increasingly experiencing frustration at the isolation and embargo in the flow of scientific and technological information from the advanced Western nations, especially from the US.

Upon the recommendation of the Science and Technology Council toward the multinationalization of science and technology, the government decided to adopt guidelines for science and technology policy 'for the first time since the end of World War II on a consensus of various ministries and agencies such as MITI and STA.'[13] The policy emphasized three points: creativity, internationalization and harmony of science and technology with human society. These guidelines were heavily ornamented with cosmetic overlay; it is difficult to put them into implementation in concrete terms. Neverthless, the policy wording could potentially be utilized in order to obtain a better slice of a pie at the time of budget allocation for each ministry within the public sector.

The cosmetic function of the slogan is for the amelioration and elimination of science and technology conflict with the USA; it was addressed both internally for Japanese science and technology as well as business communities and externally for the American political community. Internally, it still has the function of encouraging Japanese science to approach the frontline in any field by applying pressure on the Japanese science and technology community. Internally and externally, it also entailed a synergistic approach between Japanese and US science and technology communities to maintain a monopolizing position. Toward this last purpose, various agreements have been concluded between the US and Japan[14]. Science and technology issues are now an indispensable constituent of general US-Japan agreements. To mention just a few: In 1980, the U.S.–Japan Agreement for Cooperation in Research and Development in Science and Technology was signed at the presidential level. The September 1984 report, *Challenge and Opportunities in United States–Japan Relations* issued by the U.S.Japan Advisory Commissions (commonly labelled 'The Wisemen's Group') included a chapter on science and technology.

Intellectual property rights were incorporated for the first time into the 1988 U.S.–Japan S & T agreement. The 1988 U.S.Japan Science and Technology Cooperation Agreement was signed on June 20 at the Toronto Economic Summit. Cecil H. Uyehara, who has been instrumental in promoting the cooperation, recalled the 1980s, commenting that 'in many areas there has been a radical shift from the oneway street to a more balanced exchange',[15] though the response of the American industrial sector is sluggish.

Further Reduction of Governmental Power

On the area of direct aid to Japanese industry, government research subsidies have been granted since the 1950s, but they were only intended to stimulate private research; the amounts were very limited (1 percent of private spending for R&D as compared to 20 or 30 percent of US and European governments). Even these small amount of subsidies have been reduced[16] in the 1980s, as the expenditure of the private sector tends to be larger.

Also, because of the financial difficulty of the government, in contrast to the economic booming of the private sector, public laboratories are troubled by budget reductions.

During the 1980s, MITI's budget constraints have brought about the need for participating companies to shoulder a greater burden of project costs. MITI funded 40% of the VLSI Project in 1976–1980. The next project of the New Semiconductor Elements is only 25 % funded and MITI projects are increasingly privatized. Most of the joint research is being done in corporate laboratories, not MITI labs, to reduce government expense.[17]

Envious of MITI's success in the VLSI project, other government departments, such as STA and MESC, followed suit in supporting industry-related projects, but again due to financial constraints, not much in the way of achievements are expected.

Toward the Multinationalization of Science and Technology

The private sector must take a slightly different approach, although it does not explicitly state what it is really prepared to do in order to remain on good terms with the government sector, particularly with MITI.

The period of high economic growth in the 60s was MITI's heyday: it had control over the entire Japanese private sector with its authority to allocate foreign exchange and various other powers of permission and licensing. Along with the progress of liberalization, MITI has tended to loosen its bureaucratic control, while big corporations have gained sufficient power so that they no longer feel the need for the protection of MITI. MITI was also subject to international criticism of their attempt to restrict the outflow of Japanese technology, e.g. the patents originating from joint projects such as the VLSI program.[18]

Since 1980, rather than national protection, some private corporations oriented toward multinationalization, with a synergistic rather than a competitive approach to US corporations and their technology. One example is the synergism of GM and Toyota technologies, an operation which was concluded in 1983. This trend may now have the potential to monopolize the future world market.

Under the pressure of 'internationalization' MITI and, to a lesser extent, NTT (Nippon Telegraph and Telephone) give much lip service to international alliances, but concrete action is slow in coming. In spite of the reluctance of the public sector to offer concessions to international corporations,[19] there have been a number of attempts at joint ventures, some of which promote 'intermultinational' rather than international competition. Such a venture existed among several blockings of synergistic cooperation and multinational linkages in microelectronics.[20]

At this stage, however, a synergistic approach between individual companies is not so independent of a government policy line. In fact, if a company is willing to make foreign R&D investments as well as investments in foreign manufacturing facilities, that company is generally viewed favorably by foreign governments.[21] The matter of synergistic cooperation is, however, beyond the capacity of individual offices or corporations, where the mode of headlong competition is still dominant, but left to the semiformal, or informal organization of technocratic establishments of both countries, such as the United States–Japan Advisory Commission and the Advisory Council on Japan–U.S. Economic Relations, both of which entertain grave concerns about the science and technology conflict. By way of complementing each other's strong points, Gerald P. Dinneen has said, 'both countries, therefore, should work together to gain the rewards that appropriate cooperative endeavors in basic science and technology can bring.'[22] In spite of the sensational impressions to be gleaned from journalistic reports on the conflict, cooperation is quietly going on in the background.

It may be quite plausible, as John Naisbitt predicted[23] that, if harmony can be established and maintained between the United States and Japan for the coming ten years, the two countries will be a dominant force internationally through a relationship he has called 'U.S.–Japan, Inc.' As a result of this amalgamation, the rest of the world will be, technologically at least, excluded from the frontrunners and left behind. Success in this regard could seriously damage their relations with Europe and the Third World.[24]

The SDI Episode

The Reagan Administration proclaimed the adoption of SDI (Strategic Defense Initiative) in 1983 and, toward the end of that Administration, tried to involve the scientific community of Western countries to cooperate in it. This has officially opened the way for participation by Japanese scientists in US military research and development where Japan's high technology is clearly needed. The then prime minister, Nakasone, expressed willingness to send Japanese scientists to work with SDI groups if so requested by the US.[25]

While most European countries responded more or less positively, opinion within the Japanese industrial sector was divided. Heavy industries, such as Mitsubishi Juko and Komatsu Seisakusho, which have been traditionally oriented toward defence industries, welcomed the SDI proposal. On the other hand, the microelectronics industry, the real target of the American defence community, was diffident; while they would expect to acquire a valuable insight into American defense technology by cooperating and participating in the project, they entertained misgivings concerning the possibility of DOD's requirement of classification becoming incompatible with their market-oriented strategy.

Still, Japanese technology, which was originally developed for civilian use, such as the precision-required technique of ball-bearing making, has had a 'spin off' (or 'spin in' from DOD's point of view) effect into American defence technology. A survey of the United States Department of Defense (DOD) indicated that technology of potential benefit to U.S. defence systems is available in Japan and mechanisms have been established for the purpose of arranging transfers through private sector channels.[26] DOD's main interest in Japanese technology is not primarily in Japanese military technology but civilian technology, which will result in a reverse spin-off (spin-in) from civilian to military. Japanese corporations worry that their technologies' might spin-in into American military technology and then in turn be brought back (spin-off) to civilian commercial use within the States. This American civilian technology might become potential competitors with Japanese civilian technology.

In the academic and citizen sectors, JCP-oriented or otherwise, appeals and statements were issued to denounce Japanese involvement in the SDI projects and to criticise the project itself. The Physical Society of Japan has led this movement. These actions may prevent the involvement, or at least restrict the behavior, of university scientists. The project, however, was aimed almost exclusively at the private sector, over which other sectors have virtually no influence whatsoever.

Another Kind of Internationalization – Relationships with Surrounding Countries

It seems that the attention of the Japanese public and private sectors is focused on internationalization modelled on the USA; in reality, however, Japan sells a considerable amount of its technology to the NIEs contries and the Third World. While postwar American technology has been notably concentrated in defence-oriented high technology, Japan has undergone a transition from low to high technology; thus, Japan still retains some of the old technology that is no longer serviceable in contemporary Japan, and some outdated human resources, too, in the form of aged scientists and technologists who practised the technology of yesteryear. Some of them are employed by third world countries as 'silver servicemen' who guide and instruct the local workforce in the technology in the area of their expertise.

In a rapidly progressing industry such as microelectronics, complaints are often heard from countries surrounding Japan in regard to Japan's reluctance to share her technological skills compared to the case of the US. Despite such claims, the technology transfer from Japan to East and South-East Asia has been going on since the 1950s, resulting in the establishment of semiconductor manufacturing complexes in the developing countries of East Asia. This has enabled such countries to become leading suppliers to world markets at the expense of Japanese makers.[27] G. Gregory, who pioneered the transfer of electronics technology from Japan to Southeast Asia during the latter 1950s and the 1960s, concluded that:

> The arcane notion that technological genius and industrial prowess are the birthright of a select group of countries, has little relevance to the real world of the electronics age. National adversary attitudes which attempt to resist or slow the pace of change will inevitably prove counter-productive. Successful competition in this new era is not inherently a zero-sums game between winners and losers in a technological race, but is essentially a cooperative game in which nations and firms eventually benefit from all new technological progress. Moreover, those nations and firms who play the game cooperatively are most likely to reap the rewards of potential synergy which derive thereafter.[28]

Notes

1 Robert M. Rozenzweig, *The Research Universities and their Patrons* (California University Press, 1982).
2 Chalmers Johnson's *MITI and the Japanese Miracle* helped create such an overestimation, though the book itself is an excellent, well-argued, scholarly work.
3 As Tessa Morris-Suzuki put it in her *Beyond Computopia* (1988).
4 *Kikan Kuraishisu* (Quarterly Crisis) no.17 (1983), particularly my article 'Tekunonashonarizumu ron' ('On Technonationalism'), reprinted in my *Shimin no tame no kagakuron* (On Science for Grassroots People, Shakai Hyoronsha, 1984), pp.187–211.
5 Shigeru Nakayama, 'Tekunonashonarizumu ron', *Crisis*.
6 'Science in Japan', *Nature*, September 29, 1983 has emphasized this aspect.
7 Even Ezra Vogel, the author of *Japan as Number One* shares this

opinion in his letter addressed to Shigeru Nakayama, *Nihon no gijutsuryoku* pp.226–227.

8 'Japan's Inscrutable Research Budget', *Science* vol.238, p.22 (1987). One must preserve a prudent distrust of quantified data on basic science. First of all, the definition of basic science varies from one country to another and from one person to another. The university budget of the Ministry of Education is categorically considered to be the Japanese government's effort on basic science. Gary Saxonhouse criticized it on the grounds that the Japanese have included teaching time as research, and hence it must be discounted.('Technological Progress and R & D Systems in Japan and the United States', in C. Uyehara ed. *U.S.–Japan Science Exchange* (1988) p.36). In actual practice, exactly half of the university budget is considered to be research budget, for the simple reason that according to the ideal of a University it is the place where research and education merge.

9 Unpublished report, Haroto Arima, 'Kagaku Shobunya no Kenkyu Ronson Hyoka' ('Assessment of Scientific Papers'), March 1990.

10 Sheridon Tatsuno, *The Technopolis Strategy: Japan, High Technology, and the Control of the Twenty-first Century* (Brady, 1986) xiv.

11 Tatsuno, *Technopolis Strategy*, p.42.

12 Gregory, *Japanese Electronics Technology*, pp.267–268.

13 Takeyasu, p.180.

14 Hitoshi Yoshioka, 'Kagaku gijutsu seisaku ni kansuru biboroku' ('Memorandum on Science and Technology Policy'), *Kagaku shakai ningen*, no.28, (April 1, 1989).

15 Cecil H. Uyehara, 'Concluding Remarks' in Uyehara, *U.S.–Japan*, pp.256–261.

16 Takeyasu Yoshimitsu, 'Science and Technology Policy in Japan' in Cecil H. Uyehara (ed.), *U.S.—Japan Science and Technology Exchange* (1988) p.179.

17 Tatsuno, *Technopolis Strategy*, p.40.

18 William F. Finan, 'The exchange of semiconductor technology between Japan and the United States', in Cecil H. Uyehara (ed.), *Technological Exchange: the U.S.–Japanese experience*, (Proceedings of a symposium held on October 21, 1981, University Press of America), p.71.

19 Tatsuno, *Technopolis Strategy*, pp.44–45.

20 Mel Horwitch, 'The Emergence of Value-Creation Networks in Corporate Strategy', in Cecil Uyehara (ed.), *U.S.–Japan Science and Technology Exchange, Patterns of Interdependence*, (Westview Special Studies in International Economics and Business, 1988), pp.188ff.

21 Gerald P. Dinneen, 'U.S. Japanese Science and Technology: A

Comparative Assessment', in Cecil H. Uyehara (ed.), *U.S.–Japan* Science and Technology Exchange (1988) p.23.

22 *Ibid.* p.22.

23 John Naisbitt, *Megatrends* (New York, Warner Books, 1984). Its Japanese translation had been widely received by the Japanese audience

24 Justin L. Bloom, 'A New Era for U.S.–Japanese Technical Relations? A Cloudy Vision', in Uyehara, *U.S.–Japan*, p.218.

25 Toyoda, in *Assessing nuclear age* p.64.

26 J. Bloom, p.230

27 G. Gregory, *Japanese Electronics Technology* p.17 and p.9.

28 *Ibid.* pp.20–21.

10 Looking to the Future

SECTION I. Possibility of Shifting the Centre of Science to Japan

The Centre Always Shifts

Is there any possibility that the centre of scientific activities will shift to Japan? The aspect of centre shift varies depending upon the kind of science, public, private or academic. Furthermore, public policy within a national administrative boundary is not the sole determinant of centre shift, but the individual migration of scientific workers is equally a decisive factor. In this section we postulate possible trends in the future.

Scientists today, at least native English-speaking scientists, believe that most science is written and communicated in the English language. The present overwhelming usage of English practised in the world scientific community gives an impression that this state of affairs could be perpetuated in future generations as well. Such a situation could be viewed as 'English (language) Imperialism' by non-English-speaking scientists with a grievance. The latter group of scientists might be relieved, however, to find the following lesson from the history of science.

Throughout the history of science, the centre of scientific activities has continually moved from one place to another: from ancient Babylonia to classical Greece, to Hellenistic worlds, to the Islamic region, to Renaissance Europe, to seventeenth century England, to eighteenth century Paris, to nineteenth century German universities, and to twentieth century USA. If we look closely, we find that the shift has taken place not necessarily always across national boundaries. It is the language employed to express science which has changed.

Problems with Language

A national boundary of science is only a modern phenomenon, culminating today in incorporated defence science. In studies of the history of science, the language employed by the authors is more meaningful than their race or nationality. The literature reveals to us the former but not the latter. Thus what we call, for instance, Greek science, is not that practiced by the Greek race in classical antiquity, but rather science which is written in the Greek language. In this way, we can define Latin science and also Chinese science, the latter of which includes science practised by Koreans and Japanese and written in the Chinese language in the premodern period.

While Arabic science incorporated the peripheral languages of algebra and chemistry in addition to the Greek tradition, paradigms of Eastern science, however, remained unchanged up until the early twentieth century because its centre stayed in China and its scientific language continued to be Chinese. The Japanese, however, began to switch their paradigm from the Chinese to that of the West in the eighteenth century and started to translate Western scientific works.[1]

'Centre-periphery Hypothesis' in Science

When the shift of the scientific centre took place, or more precisely when the scientific language changed, the events which transpired constitute some of the major issues for students of the history of science. In the following, I shall present a hypothesis – still not fully tested – to explain shifts of scientific centre in terms of the centre and periphery dichotomy.

Scientists at the scientific centre, of course, express their profession in the central language. On the other hand, those scientists in the periphery conceive science in their native language while expressing the outcome in the central language; in other words, scientific bilinguality exists in the periphery. In the ordinary practice of normal science[2], the latter's state of affairs is considered to be disadvantageous, since peripheral scientists have to conform to the paradigm of the centre. For them, the model to follow always exists in the centre and what they must do to the best of their ability is to assimilate themselves toward that model. They can never think of surpassing the standard of the centre or of changing a given paradigm. Even if they attempted to do so, there would be neither followers nor evaluators. On the other hand, central scientists can pursue their research without being concerned as to what is going on in the periphery. (This centre-periphery relationship can also be applied to academic disciplines between basic and applied sciences.)

If the above situation continues, stagnation may result with regard to the further extension of normal scientific practice. In the centre, while scientists look for sources of information only in the central language, they may cease to be receptive to new paradigms which often emerge in the non-orthodox context of a language other than the central language. On the other hand, upon accumulation of information from the centre as well as the periphery, peripheral scientists occupy a position advantageous compared to that of central scientists due to the variety and abundance of information available to them. Chances are that out of this state of affairs, a new paradigm will emerge.

The 'threshold' conditions that prevail when the periphery becomes the centre are, first, when the language used in expressing their trade changes from the central language to their own native language, and second, when the latter begins to be cited by other language groups. If the second condition is lacking, a peripheral activity remains a local paradigm and local science, such as culture – and locality-bound sciences like some forms of geological science. Japanese social scientists read the Western language but write only in Japanese. They have more information than Western scholars and even their own local paradigm but this activity is virtually unknown outside of Japan. Only lately, as the Japanese economy attracts the attention of Western scholars, have some of the scholars started introducing Japanese works in the Western language.[3]

Events in Germany in the 1920s

In order to explain the shift of the centre from Germany to the USA in the 1920s in terms of language, I conducted a survey of citation sources of scientific articles which appeared in major physics journals such as *Annalen der Physik*, *Zeitschrift* fuer Physik and *Physical Review* in 1900 and 1925.[4] German journals cited mostly German articles (80% – 90%) in and around 1900 – although practically all German scientists could read English – whereas by about 1925, the incidence of German citations decreased to 60% – 70%, while citations from English journals simultaneously increased. American journals, throughout this period, mostly cited articles by fellow Americans, but the quantity did not exceed 40%, the remainder being citations from German and British authors.

This survey does not directly prove the validity of the above-mentioned centre-periphery hypothesis, but indications are that the Americans collected a greater variety of world-wide information, while the proud Germans, who communicated exclusively within the German-speaking scientific community around the turn of the century, apparently lost self-confidence in the course of time and cited more English journals.

Problems of Human Resources

Within the history of science discipline, there are a number of studies on the rise of scientific centres and their mechanisms, whereas there are far fewer studies detailing the reasons why the centres declined or disappeared. Not only is the latter study much less attractive, but it is difficult to approach, due to the lack of readily apparent evidence. Perhaps the normal course of development is such that, as aged science attracts less followers, it reaches a stage resembling natural death. Although the centre-periphery hypothesis might have been at work slowly from the late nineteenth century onwards[5], what occurred in Germany in the 1920s may not be the process of natural death. It must have depended also on external factors, such as the defeat in World War I and the ensuing economical hardship, finalised by Nazi intervention in the 1930s.

The state of affairs is well documented by Isidore Rabi's experience of studying in Germany in the late 1920s.[6]

In those days, German science had created its last and greatest paradigm of quantum mechanics. Rabi, who followed the American practice of studying in Germany before starting his scientific career after completing his Ph.D., was naturally attracted by the famous German scientists, such as Werner Heisenberg, Erwin Schroedinger and others. What he found in Heisenberg's seminar was that Heisenberg was the only German – more than half the seminar participants were Americans, and the rest were Chinese, Japanese and so on. The big names associated with the quantum mechanics paradigm were all there in Germany but no young followers were emerging. In his eyes, the decline of German science was all too apparent.

In those days, Rabi wrote scientific papers in German in the hope of obtaining recognition from his German peers. However, it was not language but scientific human resources, which contributed to the shift of the scientific centre. Language is only a superficial indication of centre shift, although written documents can be more easily dealt with as evidenced in historical scholarship. The relocation of scientific manpower is more essential than language change in 'migration of scientific leadership', a better expression than 'shift of centre'. A new scientific concept and new paradigm may be language-bound, but a scientific revolution can be completed only with those groups of scientists who support and develop a new concept or paradigm. In the above case, German is the original language of quantum mechanics concepts, but those concepts were developed by American followers who translated them into their own native language of English after returning home from study in Germany.

From USA to where?

Since the mid-1980s, the 'decline' theory of American civilization has been entertained among American authors.[7] Along with this argument, it may be worthwhile examining the likelihood of a future centre shift from the USA to elsewhere, despite the fact that American science remains predominant in the contemporary world. The candidate countries to be examined here as recipients are Japan in the near future, and China in the distant future.

Problems with Language

One of the requirements for a nineteenth century scientist was to be accomplished in three languages: English, German and French. This, of course, applied even to proud German *Wissenschaftler*.

Hence there were few language barriers among European scientists.

One of the Ph.D. requirements for a twentieth-century American scientist was, prior to World War II, to be accomplished in the above three languages, not only in reading but also in conversation, in order to communicate with European scientists. After the Second World War, however, language requirements were reduced to mere reading ability. Today, American scientists do not require any language other than English or, even if a language requirement does exist, it is more a matter of ceremony. Americans believe that one can practise science with English only, as major contributions to science are all written or translated into English. Scientists in non-English-speaking countries entertain some doubt in this belief, but they also know that it is largely true in the current circumstances of the scientific world.

If we apply the centre-periphery hypothesis discussed above, the present situation in American science is clearly an indication of the beginning of a declining phase, since American scientists are only able to collect information available in the central language of English. Already, some perceptive observers[8] have noticed this blunder in the American camp concerning the collection of information, and have stressed the importance of scientists and engineers learning the Japanese language.

The Centre-shift of Private Science

Over half of the scientific papers by Japanese academic scientists are now written and published in English and hence there is little problem for Western scientists in this regard. It is not, however, in this area that the Japanese show excellence, but in the field of private science, as we discussed elsewhere[9]; the fact is that Japanese academic science and university research is inferior to its American counterpart.

Academic science is evaluated by peers whilst incorporated and technocratic science is evaluated by sponsors or employers.[10] Each science has its own value system. For instance, the concept of 'creativity' differs vastly in meaning between the international scientific community and managers of big Japanese corporations. It means Nobel prizes for the former and product-marketability without international conflict for the latter.

Academic science is, in principle, public knowledge to be openly publicized and hence few problems exist with regard to ownership of intellectual property. Ownership of ideas is quite a more serious matter within incorporated and technocratic science. Hence, the mechanism of centre-shift and the centre-periphery hypothesis will operate in private science in an entirely different fashion from that of academic science.

The 'language barrier' for Japanese and other peripheral academic scientists describes the frustration and difficulty of expressing their research freely in the central language of English. The Japanese private scientists need hardly be aware of that barrier. In private science, a language barrier exists at the receiving end of information rather than at the expressing end.

Among Japanese private scientists, who have academic inclinations, a cynical dictum prevails: 'In private science, the most valuable scientific information is kept secret within the company, less valuable information is sold in the form of a patent[11], and the least valuable is openly reported in an academic journal[12].' What American scientists need is an information bank of Japanese findings in the field of private science. Such information is rarely publicized, instead stored, sometimes unwritten. If published, however, it would undoubtedly be in Japanese, appearing in an informal kind of media, such as handwritten reports or patent applications. Unlike academic science, there is little motivation on the part of Japanese private scientists to put these sources into English, unless they have a particular purpose in doing so, such as advertising in an academic science journal before selling patents abroad. Thus, it is largely left to American scientists and engineers to collect and translate this non-public and semi-public information, or even to steal it.

It is observed that the most obvious superiority of Japanese private science over its American equivalent is its capability to gather information, a process which is sometimes seen as little more than the first stage of imitating Western technology. While JICST (Japan Information Centre of Science and Technology) has monitored technical information worldwide for decades, its American counterpart NTIS (National Technical Information Service) has been unable to provide large numbers of translations of documents originally written in Japanese. NTIS plans, in the near future, to operate a computer terminal with access to an international data base of Japanese technical information that has been established by JICST. Unfortunately, the data base will have only English titles and keywords, so it will still be necessary to arrange for translations of the substantive material.[13]

The above information consists of the published material of mainly public laboratories and some academic science. Unlike academic science, which may be communicated through written publications, the transfer of private science (technology) is carried on mainly through personal contact.[14]

In terms of information flow, the central position in academic science could be maintained more easily and perhaps for a longer time compared with private science because academic information produced at the periphery is published and often written in the central language. Thus, the USA will maintain the central position in academic science in the twentieth century for some time and perhaps bring it into the twenty-first century. The language of academic science around the world will remain English, in which Japanese academic scientists will continue to write. In contrast, the language of private science has not proven to be English exclusively.[15] Hence, the centre-shift of private science could possibly shift with comparative ease and stealth to another region – to Japan or China for instance. In fact in some areas of private science, such as robotics, the centre has already shifted from the USA to Japan.

Computer Language

One of the paradigmatic sciences in postwar American science is computer science. Its language has been developed so far only in English. One might claim that, with the advent of computer translation, all scientific resources can be translated into English in the near future, and hence there would be no information shortage and no such centre shift associated with a central language change as has happened in the history of science; American science is, supposedly, forever! Up to the 1960s, I was inclined to believe in such a theory. In those days, putting Chinese ideographic characters into computers could hardly have been anticipated.

In order to make complicated characters workable, it was required to extend computer capacity far beyond the task of dealing with simple alphabets. A decade later, however, Japanese computer engineers succeeded in producing the Japanese word processors with the aid of newly developed VLSI. Larger computer capability means a greater possibility of further application and developments. The disadvantages of the complicated ideographic tradition have now turned out to be advantageous after all. It was noted by a keen Swedish observer, Jon Sigurdson that

> some experts claim that the handicap of the Japanese language has helped to improve the level for voice and handwritten input and automatic translation. The reasons given are that the problems that must be solved are more challenging than those related to Western languages. In fact, because of the early difficulty in developing an easily-used Japanese-language computer keyboard, Japanese industries have come to dominate the world market for high-speed facsimile equipment[16] to replace telex.[17] Furthermore, Japan was the first country to introduce word processors which responded to voice input. Such machines, even if slow, were already being sold in 1982.[18]

The same can be said on the matter of computer translation as well. Up to the 1960s, the possibility of computer translation was nearly abandoned because of its technical difficulty. With the advent of more capable technology, however, Japanese and Chinese engineers are now engaging in it more zealously than their American counterparts. This is merely a manifestation of peripheral psychology, but I see here the centre-periphery hypothesis at work on the matter of information collection.

Human Resources

Scientific human resources are increasingly becoming a key factor in the promotion and maintenance of a country's scientific level. Up until the middle of the nineteenth century, science was promoted and maintained by a handful of amateur scientists. The philosophical faculties of German universities in the nineteenth century devised a reproduction system of scientists, and since then the world scientific community has been progressively enlargened. A nineteenth-century Japanese scientist, Hantaro Nagaoka, observed that, whilst British and French science were maintained by a few A-class scientists, German universities had a considerable quantity of scientific human resources, in a range from A to E classes.

In the early twentieth century, American graduate schools started to produce still greater Ph.D. human resources with more systematic training. According to I. Bernard Cohen[19], while there were fewer A and B class scientists than were to be found in Europe in the prewar period, the USA produced a great number of C and D class scientists and engineers, upon whom postwar American science has been built. This is a mass science; in addition to that, the American Federal government lavished national funds on research and development to create a big science, which characterized postwar American science.

Postwar Japan, too, built up large quantities of scientific human resources from the beginning of the 1960s. In terms of sheer numbers, it is by no means equal to that of the USA; the quality differs also. Whilst the American scientific community acquired foreign scientists and students at times to suit their needs in a 'brain-drain' of other countries, this phenomenon never occurred in Japan, where scientists and students were all Japanese citizens, educated in the Japanese education system, and screened by the entrance examination of higher education. Private scientists are 'quality-controlled' in order to achieve the ambitions of corporations.

Migration of Scientific Leadership

As I stated earlier, language is merely a superficial or superstructural manifestation of centre shift. Perhaps, a more essential factor of centre shift of scientific leadership will be the migration of human resources.

Let us suppose that the scientific centre is now shifting from the USA to elsewhere. We find a considerable difference in the human resource problems of Germany in the 1920s and the USA in 1990. From the nineteenth century up to the 1920s, students of science all over the world gathered in Germany, the world scientific centre at the time. In the same way, talented students of science from countries throughout the world are now assembled in the USA for training and research. The major difference between these two migration phenomena is that while students going to Germany returned home, many current students sent to the USA stay there and become American scientists and engineers.

American Science or Asian Science?

After a decade of the science and technology boom in the post-Sputnik era, the American research and development world experienced long years of financial hardship as a result of successive Federal budget cuts in the 70s.[20] During this time, young American graduates turned their backs upon science and technology in favour of the more secure and highly-rewarded professions of law, medicine and business. To fill their places, a number of students from abroad were brought in.[21]

The research and development profession of science and technology is, in principle, a world in which race and sex are irrelevant, a world in which birth, wealth and language plays much less a role than in other professions, in other words, the world of hungry sports, where one can climb up the social ladder with one's sheer ability and painstaking labour. In the science and technology world, students with a WASP background must compete on an equal standing, and hence they choose less risky and more advantageous professions.

Foreign students from the third world seeking careers abroad tend to choose studies of science and technology, in which native language plays much less a role than in other disciplines, such as humanities and social sciences, or in other professional studies such as law and medicine.

In fact, the majority of science and engineering students in American graduate schools are now foreign-born[22]. In the trend of new American engineering Ph.D.s, the number of US. citizens has halved in the last decade and a half. Americans were outnumbered by temporary visa holders in 1984. This trend will be accelerated further, unless external measures are applied to limit enrolment of foreign students[23]. In the 1950s, the largest number of foreign scientists came from India, but the recent trend indicates that an overwhelming number originate from East Asia, particularly the so-called NIEs (Newly Industrialized Economies) countries, followed by Middle East countries[24]. 1988 statistics show that Taiwanese form the most substantial group of foreign scientists, followed by scientists from China, Korea and Hongkong[25]. One can easily predict that China will surpass Taiwan in the near future.

A new field like computer science, for instance, attracts more foreign students, as English is less of a requirement, than in other fields. In these new fields, the aged, established professors cannot compete with young graduate students in keeping up with the

Figure 14 Engineering PhDs from American institutions

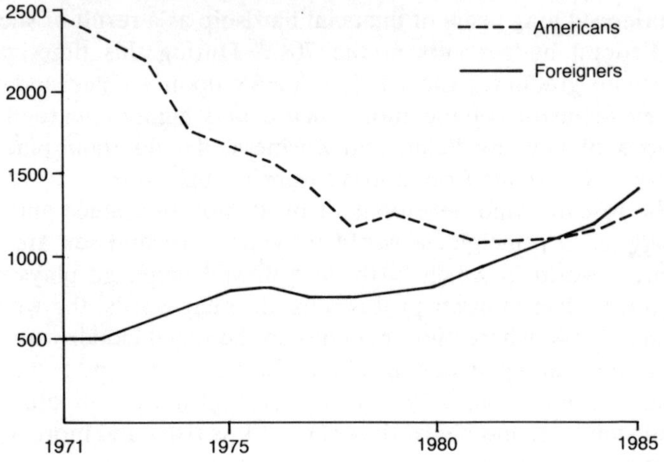

Source: *The Scientist*, Feb. 8, 1988

Figure 15 Engineering foreign students in US

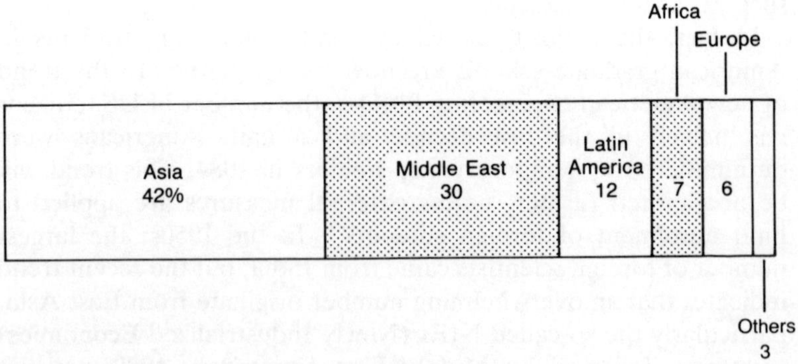

Source: *The Scientist*, Feb. 8, 1988

research frontier. Consequently, most American students taking courses in computer science are, in actual practice, instructed more often than not by foreign-born teaching assistants rather than an American-born professor who has a nominal teaching post.

Thus, it may not be too much of an exaggeration to say that in American universities in the not too distant future, foreign professors will teach science and technology to foreign students. Present statistics show that the majority of science and technology Ph.D. holders are foreign-born and more than 60% of foreign students do not return home after the completion of their education, preferring to pursue their careers in the USA. The numbers of Chinese students who remain exceeds this percentage.

Can the above phenomenon be properly labelled a shift of scientific centre from the USA to elsewhere: Asia, notably China? It should not be. It can be said that American society is supported by the vital energy of immigrants who infiltrate the lower strata of the society and try to move upward. As long as American society is open and generous to foreign students, the world's scientific centre will remain there forever.

Instead of learning foreign languages, Americans supplement their shortage of scientific information by absorbing scientific human resources from abroad into their scientific community.

I have so far maintained that the migration of human resources is a more important factor than the indication of language change in centre-shift. This is more definite in private science than academic. In the place where no such open communication is guaranteed as in private science, the most effective method of information gathering is the relocation of human resources with esoteric types of information installed in them. Just as a company scouts a competitor's human resources to enrich their information pool, peripheral human resources will bring with them information already existent in the periphery to enrich the open American centre.

Japan: Stagnation?

It is premature to say that Japan will become a world centre of scientific activity comparable to the American one. Japanese science has, however, established certain distinct characteristics during its postwar development; its origin goes back, indeed, to the late nineteenth century. The transfer of technology was closely associated with European migration and settlement pattern first to the United States and then to other parts of the world. A remarkable exception has been Japan[26], where transfer has been carried out by the Japanese. The Japanese have been successful in collecting technical information abroad without having accepted the migration of Western settlers.

Thus, the Japanese infrastructure of transfer is quite different from the rest of the world. Japanese science is characterized by its secluded nature and the Japanese scientific community is secluded linguistically from foreign scientists. Along with linguistic seclusion, other systems like those of universities and laboratories are designed only for the Japanese, or Japanese-speaking people. While Americans believe that competition and hybridization among different modes of thought is indispensable in promoting and generating scientific paradigms, the characteristics of Japanese private science are known as inhouse co-operation and co-ordination of researchers within a secluded system. These researchers are 'quality-controlled' through a homogeneous education and selection system.

To what extent and for how long these characteristics can be maintained in future is open to question, as Japan has recruitment problems for the next generation of scientists and engineers. If the present state of affairs continues, Japan's foremost technology of micro-electronics will be 'embargoed' by American-educated Asian scientists who are deployed in the USA as well as NIEs countries – 'Pacific Rim embargo', one might say – and isolated from the rest of the world. A Japanese micro-electronic engineer boasted that

> in the field of science and technology, adding together second-rank scientists and technologists of the USA and NIEs countries is meaningless; it is not a matter of quantity but quality. Only top-notch technology like ours can survive in market competition.[27]

This statement may be true at this moment, but in view of the incessantly changing nature of the research frontier of science and technology, for what period they can maintain such a lead remains to be seen. Even if Japanese microelectronic technology could maintain the status of world centre for some time, a secluded language group has limitations of its own; soon the centre-periphery hypothesis will take effect and Japanese hegemony will prove to be short-lived.

Qualification of the central position is not only a matter of excellence in the quality of science and technology but also depends on the infrastructure of the education communcation system. In the nineteenth century, German universities became the world centre for foreign students to visit whilst the equally prestigious French *grands ecoles* were not. This is partly because German universities conferred Ph.D.s on foreign students. In the twentieth century, American graduate schools also attracted a great number of foreign students, because they were generous in giving financial assistance and conferring Ph.D. degrees on foreign students, a practice which continues today. Most Chinese students choose the USA for their study location, mainly because of the fellowships available. It was recently recognized among science policy-makers that the American system comprises not only the generous givers, but, at the same time, the receivers of information and human resources from the periphery[28].

Japanese institutions are still far inferior to their American counterparts in providing an infrastructure for foreign students, despite recent governmental slogans of 'internationalization'. There has been a recent tendency in Japanese graduate schools to accept students from the third world, the number amounting to one third of engineering doctorate candidates in Japanese universities[29], but impoverished Japanese university research may not attract many foreign students compared to American universities or elsewhere. Foreign students prefer to be trained in much more prosperous, presently world-renowned, industrial laboratories. Private science, however, by its nature, is not generally accessible to outsiders.

Furthermore, parallel to the experience of the USA since the 1970s and Europe since still earlier times, a tendency can be seen among Japanese to stay away from science and technology in their career selection, in preference to other professions of law, business and medicine. In the late 1950s and early 1960s when Japan achieved high economic growth, science and technology attracted the best students. Since then, however, science has become less appealing, except for some limited fields such as electronic technology. Students presently believe that science and technology require hard training but that it is ultimately less rewarding in terms of career prospects compared to other disciplines. Even new graduates with science and engineering training are tending to stay out of the manufacturing industry, particularly since 1987.[30] This trend may be unavoidable in any affluent society. In Japan, the maintenance of the excellence of Japanese private science will depend on how effectively Japan can attract and exploit the scientific human resources of surrounding Asian countries.

For the past decades Japan has created its own unique infrastructure of private science as discussed in chapter 4; it will not easily collapse, but continue to function for the next generation to come. Technological imperatives force Japan to keep running on the incessantly information-creative capitalistic track. But who is running? Scientists and engineers of course. If there is a hole in the solid Japanese infrastructure, it is the certain future crisis of insufficient scientific manpower.

For the immediately foreseeable future, there will be competition between US-Asia manpower multinationals and Japan-Asia manpower multinationals. Those who are successful in thorough multinationalisation in the above triangular relationship will be the only survivors.

Thus, looking towards internationalisation in the near future, Japanese science may well remain isolated from the rest of the world and in turn, stagnate.

SECTION II. Assessment of the Privatization of Science

In the present situation of the dominance of private science in Japan, the most urgent and crucial problem is how to control privatization of science and to enhance scientific services for the citizen sector. We shall conclude this book by suggesting the direction in which we should head, if not a concrete strategy.

Who can control privatization of science?

The research frontier of industrial science moves forward incessantly as a necessary prerequisite for information capitalism to survive and continue. One might naturally wonder what will be the ultimate consequence of the technology imperative in a fiercely competitive environment. Unless the outsiders, public, academic or citizen sectors, are able to effectively set a limit on growth of information, one cannot escape from this overwhelmingly uneasy feeling. Furthermore, it is the privatized feature of science that information capitalism seeks. Japanese science has the conspicuous feature of the dominance of private science over other kinds of science, public, academic and service. Toward the turn of the century, this trend will be accelerated to the extent that nobody else will be able to control it.

From the citizen sector's view point, it would be obviously desirable to minimize privatization of science so that scientific information can be enjoyed and utilized by all. Who, then, is qualified to carry out the job to turn privatized information into public knowledge?

Traditional rights of scientists to publish research results are usually denied because of the policy of giving priority to the interests of the business establishment. Especially for those scientists and engineers in industrial science, the rights to disclose corporate secrets are categorically denied, even in cases where scientific information is of a crucial nature for human benefit. I cannot be too optimistic about the Ombudsman function of the scientific community. Only a small portion of the community is truly concerned and qualified as guardians of the interests of citizens, the rest being prone to undertake whatever research is best funded. Most contemporary scientists are involved with industrial science, the collective inclination of the community is not necessarily sound. Since we cannot expect much from academic science as a counterbalance to the overdominance of industrial science, we must inevitably turn to another kind of enlightenment, that of service science. We may owe science journalism, which will prevent overprivatization of science and expose them for the citizen's own critical judgement and choice.

Journalists present serious problems but never solve them. Journalistic provocations are relatively short-lived. Before society tires of repeated alarms, someone must take a step toward solving the problems. It is the official duty of the non-military sector of public sciences to undertake such service science. The daily life of taxpayers is conducted not on the basis of competition but rather compassion and cooperation, and so also should service science be conducted for the betterment of ordinary life. However, in the absence of such competition as has existed in the academic and industrialized sciences, service science for the people's sake lacks adequate financial support and competitive motivation and so remains insignificant and powerless. The infrastructure building of publicly-sponsored service science as a countermeasure of privatization of science should be promoted. There are some attempts made at a local government level to organize a board of citizen assessment of public science.

Questioning our technological future – a long term problem

Is there any intrinsic limit to the growth of scientific information in information capitalism? There must be a human physiological limit for the acquisition of information. Before reaching that point, however, mankind would be sharply divided into two camps; the winners and the losers in an information society; namely those who have an interest in promoting information growth and those who do not.

Quality Technology

'Japan is a winner in the contemporary competition for unerring precision. Japanese quality control is excellent; Japanese engineers pursue 100% precision while Americans are satisfied with 70%.' The above is typical of the self-congratulatory utterances of Japanese technonationalists.

Modern information capitalism exploits and appreciates a special kind of human ability, the ability to be able to assimilate oneself to mechanical infallibility. However, most of the human race may not be qualified to live up to and survive in such a precision-oriented society. Is precision technology truly the technology of quality? From a technocratic point-of-view, yes. But from the humanistic view-point of the citizen sector, we may say on the contrary that the technologies which do not permit the presence of human errors are poor and inferior in quality.

A technocratic spokesperson claims that the goal of Japanese exact science and technology is the creation of a society in which people are able to enjoy a carefree, leisurely lifestyle, liberated from every kind of precision work which can be left, instead, to robots and computers. A question still remains, however. Such a society may still contain a hierarchy of classes just like the societies before it. Those who are able to utilize future science and technology are technocratic elites. They design a precision-oriented society in which most people will be subordinate to them in a slavelike position. Such a polarization is also found in the North-South dichotomy, in which science and technology plays a major role in increasing the gap and difference.

The Limits of Competition

Science is promoted within a competition mechanism, whether individual, inter firm or interstate. Like bicycle-riding, information capitalism is kept going with a drive incessantly compelled by inter-firm competition.

As a successful competition mechanism Japanese secondary education is internationally praised for its high degree of mathematical achievement. Such mathematical ability in future scientific workers is evaluated as an imperative in an information capitalism. However, in Western history, school mathematics exams were introduced mainly because they were the best way of competitive examination. They presuppose a single answer to a single question, no matter whether it has any correlation with a scientific career. The exams also require a rigorous regimented training of problem-solving, rather than problem-creation. It is the best test of one's ability to endure harsh competition and regimentation. It is this aspect of mathematical achievements which the Japanese educational establishment is aiming at. Thus, the Japanese world of industrial science and the world of examination-oriented education has a common competitive rat-race strategy. This is the infrastructural environment of Japanese information capitalism.

Such a competition-driven society is fragile and vulnerable. Technocratic leaders are all firm believers in a dictum of 'competitionism': competition bringing about human progress. The dictum is a time-honoured truth but it is also true that the dictum can be welcome only to the winners of competition. Furthermore, it is also true that one cannot be a winner all the time. Competitionism is the guiding principle of public, private and academic sciences but not that of the citizens, whose everyday life-styles are performed without the drive of competition. Some time in the future,

Japanese citizens will be freed from the possession and obsession of competitionism.

Pollution of the Intellectual Environment

Suppose that the maldistribution of information is modified to the extent that nobody suffers any disadvantages from inaccessibility to information resources. Still a fundamental problem remains unsolved once they realize that competition is not necessarily compatible with the quality of life.

The average citizen, who has no vested interest in science and technology, always remains comparatively more the 'loser' in a technocratic society. The difference between a 'loser' and a 'winner' is ever increasing as science develops. The loser never has a chance to catch up and decrease this difference, which invites discrimination.

Nuclear establishments view community anxiety regarding reactors as being due to the people's ignorance of science and technology. The anxiety can never be banished as long as the disparity of information on reactors exists between the establishment and residents. And the disparity can never be banished in principle, because residents have no time to catch up with the processing of information which incorporated scientists produce every day. Those who suffer from this disparity may call it intellectual pollution[31] or intellectual imperialism.

In our present information society, the disparity of information means discrimination. Unless it is established as a human right that nobody is discriminated against unfavourably even though they are ignorant of certain scientific knowledge, residents in future societies will increasingly be tormented.[32] Ultimately, the most unique problems inherent in 'science for democracy' is the matter of human rights of ignorance on scientific matters. It sounds like a paradox but it certainly turns out to be more essential as long as science and technology continue to develop further.

Having human rights of ignorance established, the decision that people are concerned is then only an ideological, not a scientific matter. Human rights legislation can free and protect people from the overwhelming pressure of piecemeal knowledge, which technocratic scientists produce day and night. This consideration may be eventually represented by parliamentary procedure; then, the citizenry will be able to enjoy their free time in whatever pusuit they desire, without being pressed upon by the piecemeal knowledge of science such as the knowledge of atomic reactors and recombinant DNA experiments which have no relevance to his or her enjoyment of life. Only in this way, will science and technology be welcomed by all of mankind, devoid of technocratic intimidation.

Science for Everyday Life – A Concluding Remark

We have so far discussed science, technology and society in relation to a nation state, Japan. However, as long as we are tied to the interest of a nation state, we may not be able to reach any solution to contemporary problems. At this point, we should turn to ourselves and start questioning what science and technology means to us personally. We must ponder the following questions. What sort of life-style would we like to follow? What sort of science and technology, what resources are needed for the life-style we prefer?

For prewar Japanese, science and technology are not primarily for the life of the people. They are primarily for the purposes of armament. In the postwar period, freed from the burden of armament, the major target of science and technology promotion was set for economic recovery and some part of their resources are turned to the improvement of people's life-styles. In the early 1980s, a slogan of techno-nationalism has appeared. With this slogan, we are immediately reminded of the prewar pattern of research and development, namely science for enriching the nation, strengthening armaments and consequently impoverishing people's lives again.

Even in the postwar period, the Japanese have not opted to use their science and technology for the betterment of their own lives. First, corporations set the goal of R&D and carry them out. Only after corporations disperse their products into market and environment, do people adjust themselves to the new commodity environment. People are obsessed with the belief that, if they fail to adjust themselves, they may not be able to survive in an information society. Concerned about the slogan of 'microelectronic revolution' and fearful of deskilling, aged workers attend computer courses after office hours. If all the Japanese further adjust themselves to technonationalism out of a belief that it is the trend of tomorrow, what will be the outcome? The result will probably be that the Japanese economy will be even stronger and thus become the source of further international conflict.

It is often commented that contemporary Japanese R&D lacks a clearly set goal apart from profit-making. At the strategy level, there are two major ways of promoting science and technology. One is science for the state and corporation. To promote the wealth of a nation, corporations must flourish. For that end, we have to develop science and technology. In this way, it will eventually lead to the improvement of the living standards of the populace. The above is the logic of technonationalism. The second way is a more scientistic one. No matter whether or not we know how to make use of them, we have to promote the fifth generation computer or DNA recombinant experiments, as they are scientifically feasible. The research results will be utilized by corporations and the state. Eventually it will also benefit the Japanese people. This logic of technological push is often followed at the research frontier of high technology. Neither of them make a priority of the benefits to the citizen sector, which is instead placed at the incidental end of total operation. Those who have no vested interest in the promotion of science and technology now ought to question whether the conventional priority is a valid one.

Right after the war, many hungry and devastated Japanese hoped that science and technology might bring forth a miraculous recovery of Japanese living standards. The desired effect was not readily brought about. The technocrats in the public sector planned and adopted a recovery policy of 'differentiated production,' in which basic industries, such as energy and steel, were highly supported and subsidized by the government at the expense of the production of consumer goods. Japanese people have endured hardships for a long time. Only as late as late-1950s and early 60s, has the living standard of the people visibly improved. Nevertheless, that postwar hungry generation still hope and firmly believe that science and technology will further improve the material life of the people.

However, what the people expect to derive as benefits of science and technology must have undergone major changes in the early '60s. Technocratic establishments did not take notice of this change and continued the line of development during the hungry days of the 1940s and early '50s into the late '60s, during which the positive value added by old science and technology decreases and on the contrary negative elements increased. Thus, the accumulated tension of the society was suddenly released in the form of a grassroots revolt in the early '70s, as evidenced by the anti-pollution movement and even anti-science movement.

In the meantime, the standard of living, and, indeed, the environment in which they live, have been largely transformed. Though the value of science and technology appears to be unchanged, what people are looking for in science and technology is subject to change as the circumstances of their lives change. What was desired at the subsistence level in the 1950s is qualitatively different from that which is hoped for now in the 80s and '90s. Failure to recognise these changes and the thoughtless extension of old value systems into the twenty first century will undoubtedly lead to catastrophe. MITI in the 1970s projected a seven percent rise in energy consumption each year by extrapolating a simple line which described the trend of the high economic growth period. The Ministry accordingly designed an energy supply system largely dependent upon nuclear energy. Does this mean that the temperature of the houses of citizens goes up seven percent each year? Does it also mean that people extend their distance of travel seven percent each year? People would like to set a limit to such growth.

The Japanese people have so far sacrificed their own interests for the benefit of the state and corporations on the basis of wishful thinking that such benefits will, eventually, be shared by the people. Compared to the prewar period when priority given to military science and technology did not provide any returns to the people, the postwar priority of private science has, to a certain extent, contributed to the benefit of the people, but not enough. This is the time when the citizen sector start to examine their own life-style and work-style prior to the interests of the state and corporation, establish their ideal ones and only after that select the brand of science and technology that they truly wish to have and speak up. In other words, instead of adjusting their lifestyles to the public and private sciences provided by the state and corporations, citizens have to find out how to obtain the service science that they truly need and want. Recent white papers on Japanese science and technology are aimed at improving the life-style of the Japanese people. Though the government has not provided any clues as to how they should go about it, citizens should be more outspoken in making known what they as individuals would like to have.

Notes

1 Shigeru Nakayama *Academic and scientific traditions in China, Japan and the West* (University of Tokyo Press, 1984) and also his 'The transplantation of modern science in Japan' *Center for Studies in Higher Education, the University of California, Occasional Paper,* no.23 (1982), 26p.

2 In the sense of Thomas Kuhn *Structure of Scientific Revolutions* (Chicago U.P., 1962).

3 One recent interesting example is Tessa Morris-Suzuki, *A history of Janpanese Economic Thought*, (Routledge, 1989).

4 Shigeru Nakayama, 'Kagaku no chushin no ido to paradaimu' ('Shifts of scientific centres and paradigms'), *Kanagawa daigaku hyoron*, no.4, (1988), pp.14–21.

5 Critical arguments of Friedrich Paulsen on German university research at the end of the nineteenth century and the famous memorandam of Adolf von Harnack at the time of founding Kaiser-Willhelm Gesellschaft before World War I, are indications of German crisis-consciousness.

6 Rabi's address to the History of Science Society, annual meeting at New York in December 1957. Also John S. Rigden, *Rabi, Scientist and Citizen* NY. Basic Book, (1987).

7 Ira C. Magaziner & Robert B. Reich, *Minding America's Business; the Decline and Rise of the American Economy* (1982) is one of the early examples.

8 Such as Senator John D. Rockefeller.

9 Shigeru Nakayama 'Independence and choice: Western impacts on Japanese higher education' *Higher Education* 18 (1989) p.46. A blunt statement by a non-academic observer simplified the matter clearly, 'The Japanese can't compete (with the United States at this [basic science] level) because their universities are not research organizations; they concentrate on education and training, while the research is conducted in the corporation.' in Robert F. Johnston and Christopher G. Edwards, *Enterpreneurial Science; New Links between Corporations, Universities and Government* (Quorum Books, New York, 1987) p.75.

10 For definitions of the terms 'academic science' and 'industrialized science', see Jerome Ravetz *Scientific Knowledge and its Social Problems* (1971).

11 'The Munich based IFO Economics Institute argues that patent regis-

trations in Japan help in worker motivation and public relations, and that they are often made in order to avoid disputes.' From Ron Matthews, 'Technological Dynamism in India and Japan' in Erik Baark and Andrew Jamison (eds.), *Technological Development in China, India and Japan*, (MacMillan, 1986) p.175.

12 Publication in an open academic journal is employed as a means of public relations before the company sells the patent.

13 Justin L. Bloom, 'A New Era for U.S. Japanese Technical Relations? A Cloudy Vision' in Cecil H. Uyehara ed. *U.S. Japan Science and Technology Exchange: Patterns of Interdependence* (1988) pp.228–229.

In this regard, I recall that in the late 1950s, I was requested by the Library of Congress to conduct a project to make Japanese scientific works available in English. They remarked at the time, the postSputnik boom of American science and technology, that they were in possession of sufficient funds to enable the translation project to go ahead. Translators, rather than funding, however, proved to be the problem, and the project was not able to proceed. Later, I heard complaints that they only translated titles of journal articles, a virtually pointless pursuit. Three decades later, the US are taking this project more seriously under a financial constraint.

14 OECD, *The Conditions for Success in Technological Innovation* (Paris 1971) p.98.

15 Sheridan Tatsuno, *The Technopolis Strategy* (1986) xii.

16 It has been proved to be much more convenient and useful for sending ideographic letters than alphabetical telex.

17 In Japan, the use of international telex reached its peak in 1984, international telephone calls skyrocketting since then. *Seishonen Hakusho* (1988) p.13.

18 Jon Sigurdson, "The High Technology Challenge and Policies in Japan and Sweden", in Erik Baaek and Andrew Jamison (eds.), *Technological Developement in China, India and Japan* (Macmillan, 1986) p. 63.

19 Paraphrased from a lecture delivered at Harvard University entitled, 'Science in America', in 1955.

20 B. L. R. Smith and J. J. Karlesky: *The State of Academic Science* vol.1, (Change Magazine Press, 1977).

21 Statistics illustrating this trend can be found in *Science Indicators 1982* (Washington DC) p.28.

22 Foreign students now receive the majority of engineering doctoral degrees awarded annually by U.S.universities. Sources: unpublished tabulations from the National Research Council's Doctorate Records file quoted in *The Scientists Feb.*8, 1988.

23 This measure is actually now taking place in some graduate schools in state universities, perhaps because of obligations to taxpayers.
24 Foreign-born graduate engineers in the United States are most likely to come from Asia and the Middle East. Sources: Institute for International Education, Profiles, 1983/84, quoted in *The Scientists* Feb.8, 1988.
25 *The Scientists*, Feb.8, 1988.
26 Nathan Rosenberg, *Inside the Black Box: Technology and Economy* (Cambridge, 1982) p.249.
27 NHK TV interview (1988). This is acknowledged by a 1987 Department of Defense report, stating that, of 25 microelectronic technologies, Japan led or was gaining the technological lead in 19; in 5 there was Japanese–U.S. parity; and in only 1 was the United States maintaining a technological lead. Glenn J. McLoughlin, 'Technology policy in Japan and the United States', *CRS Review* July 1989, p.8.
28 Shigeru Nakayama, *Amerika daigaku e no tabi* (Trip to American Universities, 1988).
29 'Rikakei Daigakuin' p.11.
30 *Kagaku gijutsu hakusho* (White paper on science and technology, ed. by Science and Technology Agency, 1988), p.109.
31 Shigeru Nakayama, 'Ochikobore no kagakuron' ('Science for the dropouts'), *Shimin no tame no kagakuron* (Science studies for citizens, 1984) p.15.
32 C.G. Weeramantry (ed.), *Human rights and scientific and technological development*, (forthcoming).

INDEX

Note: Most references are to Japan, unless otherwise specified